北京课工场教育科技有限公司 出品

新技术技能人才培养系列教程

大数据核心技术系列

Hadoop 数据仓库实战

肖睿 兰伟 廖春琼 / 主编

刘兵 李永明 胡淑新 / 副主编

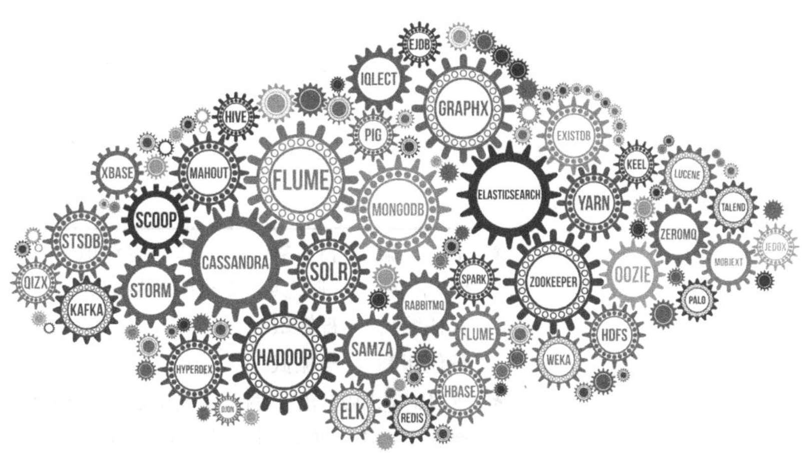

人民邮电出版社

北京

图书在版编目（CIP）数据

Hadoop数据仓库实战 / 肖睿，兰伟，廖春琼主编
. -- 北京 : 人民邮电出版社，2020.1（2024.2重印）
新技术技能人才培养系列教程
ISBN 978-7-115-52609-0

Ⅰ. ①H… Ⅱ. ①肖… ②兰… ③廖… Ⅲ. ①数据处理软件—教材 Ⅳ. ①TP274

中国版本图书馆CIP数据核字(2019)第252423号

内容提要

本书以 Hive 为开发平台，主要介绍了如何使用 HiveQL 来查询和分析存储在 Hadoop 分布式文件系统上的大数据集合，具体内容包括 Hive 入门、Hive 数据库及表操作、Hive 元数据、Hive 高级操作、Hive 函数与 Streaming、Hive 视图与索引、Hive 调优、Hive 与 HBase 集成、数据迁移框架 Sqoop 等。本书介绍的每个任务都运用了大量案例，紧密结合实际应用，融入了含金量十足的开发经验。在此基础上，本书通过丰富的练习和操作实践，帮助读者巩固所学的内容。本书配以多元的学习资源和支持服务，包括视频、案例素材、学习社区等，为读者提供全方位的学习体验。

本书适合作为计算机、大数据等相关专业的教材，也适合具有一定 Linux 或 Java 开发基础且想从事大数据开发的人员阅读学习，还可以作为大数据分析与运维人员的参考用书。

◆ 主　　编　肖　睿　兰　伟　廖春琼
副 主 编　刘　兵　李永明　胡淑新
责任编辑　祝智敏
责任印制　马振武

◆ 人民邮电出版社出版发行　　北京市丰台区成寿寺路 11 号
邮编 100164　　电子邮件 315@ptpress.com.cn
网址 http://www.ptpress.com.cn
固安县铭成印刷有限公司印刷

◆ 开本：787×1092 1/16
印张：16.25　　2020 年 1 月第 1 版
字数：350 千字　　2024 年 2 月河北第 9 次印刷

定价：52.00 元

读者服务热线：(010)81055256　印装质量热线：(010)81055316
反盗版热线：(010)81055315
广告经营许可证：京东市监广登字20170147号

大数据核心技术系列

编 委 会

序　言

知名高管语录

人类正从 IT 时代走向 DT 时代。未来的制造业要的最大的能源不是石油，而是数据。

——阿里巴巴集团董事局前主席马云

“互联网+”的发展与大数据、云计算密不可分。

——腾讯公司董事会主席兼 CEO 马化腾

探索数据的价值、挖掘大数据时代的商业模式，是全行业的当务之急。

——小米公司董事长兼 CEO 雷军

丛书设计背景

当你在知名人士口中听到“大数据”时，其实它早已渗透到了每个行业和业务职能领域，并成为了重要的生产因素。企业利用大数据贴近用户、加强业务中的薄弱环节、规范生产架构和策略，创造了更多的商业价值，进而形成了包括大数据采集、存储、处理、分析、可视化呈现等的大数据产业，并在其形成过程中提出了以 Hadoop 为代表的一整套大数据技术解决方案。

大数据产业当前仍处于技术高速发展时期，需要使用到很多不同的框架和工具，初学者在学习时会有无从下手的感觉，因此“大数据核心技术系列”丛书应时而生。该丛书根据企业的实际人才需求，参考历史学习难度曲线，选取了“Hadoop+Spark+Python”技术集作为核心学习路径。编委会系统打造大数据核心技术系列丛书，旨在为读者提供一站式实战型大数据开发学习指导，帮助读者踏上由入门到实战的大数据开发之旅！

丛书核心技术

“大数据核心技术系列”丛书以 Hadoop、Spark、Python 三个技术为核心，根据各个技术的不同特点，解决在大数据离线批处理和实时计算两个过程中所遇到的问题。主要内容如下：

- 以 Hadoop 为核心完成大数据分布式存储与离线计算；
- 使用 Hadoop 生态圈中的日志收集、任务调度、消息队列、数据仓库、可视化 UI 等子系统完成大数据应用系统架构设计；
- 使用 Spark Streaming 和 Flink 实现大数据的实时计算；

- 使用基于 Python 的 Scrapy 爬虫框架实现数据采集；
- 使用 NumPy、Pandas 和 Matplotlib 完成数据的分析与可视化；
- 使用 Scala 实现交互式查询分析与 Spark 应用开发；
- 结合大量项目案例完成大数据处理业务场景的实战。

丛书特点

1. **逆向课程设计**

满足企业对人才的技能需求是设计本系列丛书的核心原则，为此，课工场大数据开发教研团队采用逆向课程设计法（对应的设计流程如下图所示），不断迭代优化课程，形成了落地生根的应用型人才培养体系。

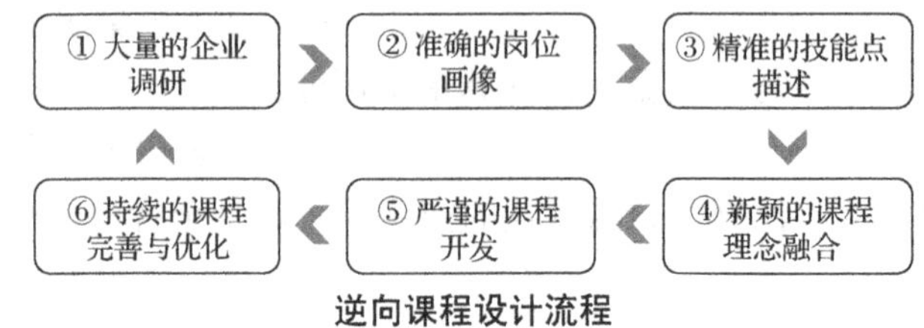

逆向课程设计流程

2. **任务驱动讲解**

本丛书中的技能点和知识点均由任务驱动，读者在学习知识时不仅可以知其然，还可以知其所以然，有助于读者融会贯通、举一反三。

3. **实战技术提升**

本丛书均设置项目实战环节，综合运用书中的知识点，帮助读者提升项目实践能力。每个实战项目都设有相应的项目思路指导、重难点讲解、实现步骤总结和知识点梳理。

4. **融媒体移动学习**

本丛书可配合使用课工场 App 进行移动学习，观看理论讲解和案例操作的配套视频，同时课工场在线开辟教材配套版块提供案例素材及代码下载服务。此外，课工场还为读者提供了体系化的学习路径、丰富的在线学习资源和活跃的学习社区，方便读者随时学习。

读者服务

读者可以扫描右侧二维码访问课工场在线的系列课程和免费资源，如果学习过程中有任何疑问，也欢迎发送邮件到 ke@kgc.cn，我们的课代表将竭诚为您服务。

感谢您阅读本丛书，希望本丛书能成为您大数据开发之旅的好伙伴！

“大数据核心技术系列”丛书编委会

前　言

大数据核心技术系列教材是为大数据技术学习者量身打造的学习用书，可实现对大数据领域核心技能的全面覆盖。

本书面向对大数据技术感兴趣的学习者，旨在帮助读者理解 Hive 的工作原理，掌握使用 Hive 进行大数据处理与分析的能力。

本书的写作背景

数据仓库是为企业所有级别的决策制定过程，并提供所有类型数据支持的战略集合，是企业大数据系统的重要组成部分。Hive 是目前企业中使用很广泛的数据仓库工具。开发人员在 Hive 中可以通过使用类 SQL 语句实现 MapReduce 快速统计，进而提高大数据开发效率。

如今 Hive 应用已经成为了大数据开发人员必须要掌握的核心技能之一。本书将全方位讲解 Hadoop 数据仓库解决方案 Hive 的应用方法，并通过提供丰富的案例、练习和项目，提炼大量的业务需求，强化读者的项目实战能力。

Hive 学习路线图

为了帮助读者快速了解本书的知识结构，我们整理了本书的学习路线图，如下所示。

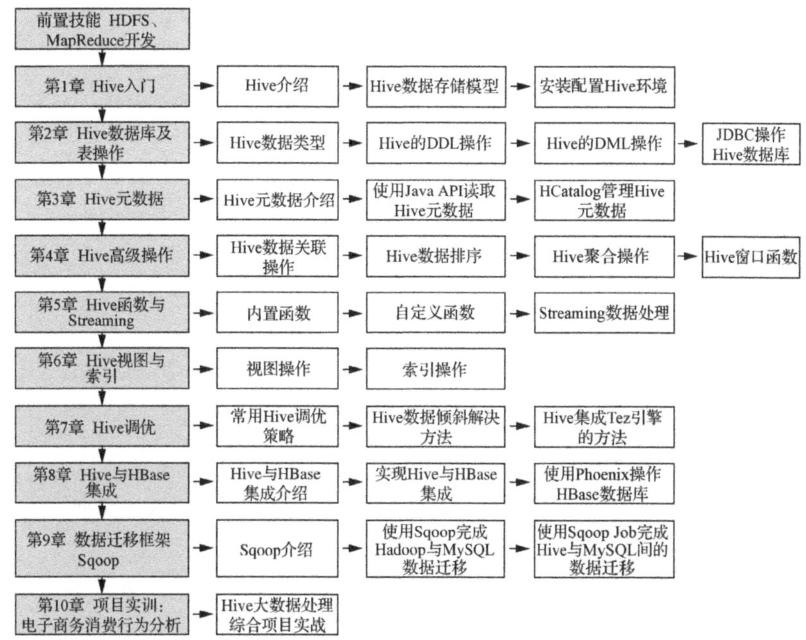

本书特色

1. **贯穿项目与实战项目相结合**

贯穿项目：雇员信息、零售数据分析贯穿每章，及时巩固 Hive 各知识点。

实战项目：电商消费数据分析综合实训，提升 Hive 企业级应用开发能力。

2. **教学资源丰富多样**

配套素材及示例代码。

每章课后作业及答案。

重难点内容讲解视频（扫码直接观看）。

3. **学习效果随时可测**

每章提供“本章目标”及“重难点”，助力读者确定学习要点。

课后作业辅助读者巩固阶段性学习内容。

课工场题库助力在线测试。

读者对象

大数据技术相关从业人员。

大数据技术的相关爱好者和自学者。

各类高校及培训机构中大数据相关专业的教师和学生。

本书由课工场大数据开发教研团队组织编写，参与编写的还有兰伟、廖春琼、刘兵、李永明、胡淑新等院校老师。尽管编者在写作过程中力求准确、完善，但书中不妥之处仍在所难免，殷切希望广大读者批评指正！

智慧教材使用方法

扫一扫查看视频介绍

由课工场“大数据、云计算、全栈开发、互联网 UI 设计、互联网营销”等教研团队编写的系列教材，配合课工场 App 及在线平台的技术内容更新快、教学内容丰富、教学服务反馈及时等特点，结合二维码、在线社区、教材平台等多种信息化资源获取方式，形成独特的“互联网+”形态——智慧教材。

智慧教材为读者提供专业的学习路径规划和引导，读者还可体验在线视频学习指导，按如下步骤操作可以获取案例代码、作业素材及答案、项目源码、技术文档等教材配套资源。

1．**下载并安装课工场** App

（1）方式一：访问网址 www.ekgc.cn/app，根据手机系统选择对应课工场 App 安装，如图 1 所示。

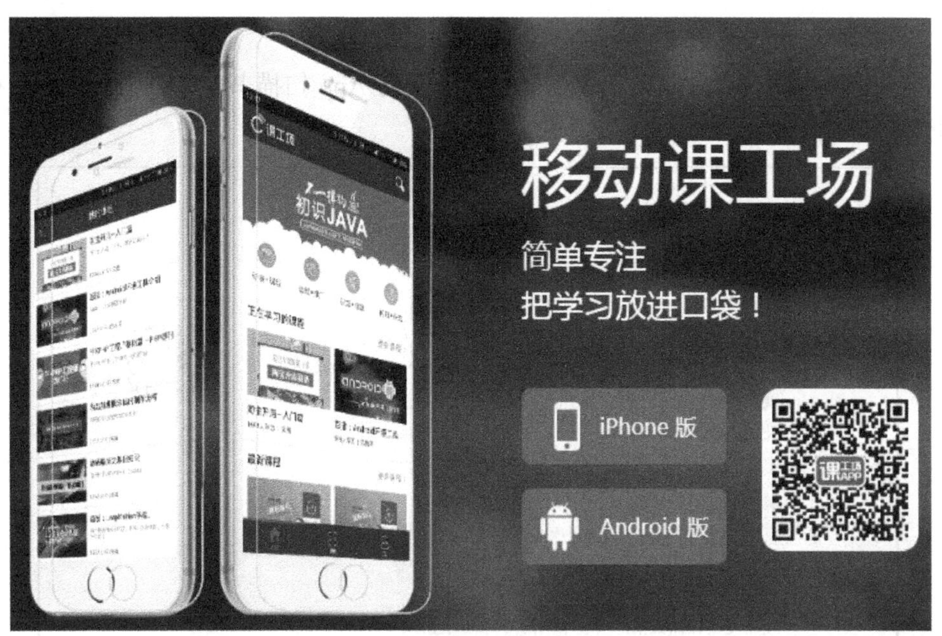

图1　课工场App

（2）方式二：在手机应用商店中搜索“课工场”，下载并安装对应 App，如图 2、图 3 所示。

图2　iPhone版手机应用下载

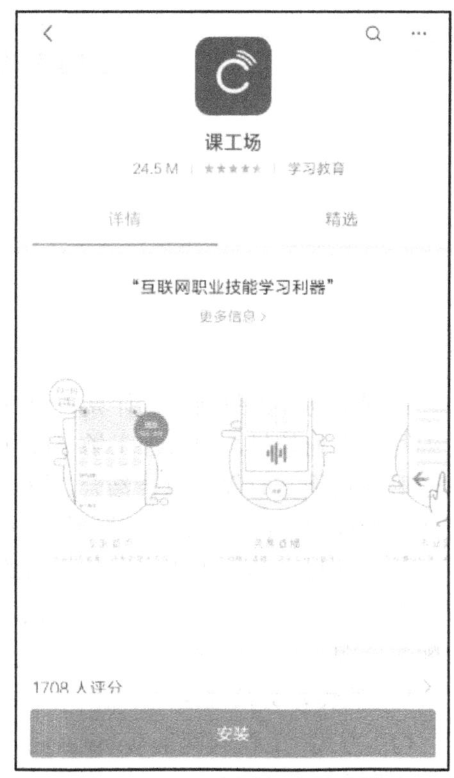

图3　Android版手机应用下载

2. ***获取教材配套资源***

登录课工场 App，注册个人账号，使用课工场 App 扫描书中二维码，获取教材配套资源，依照图 4 至图 6 所示的步骤操作即可。

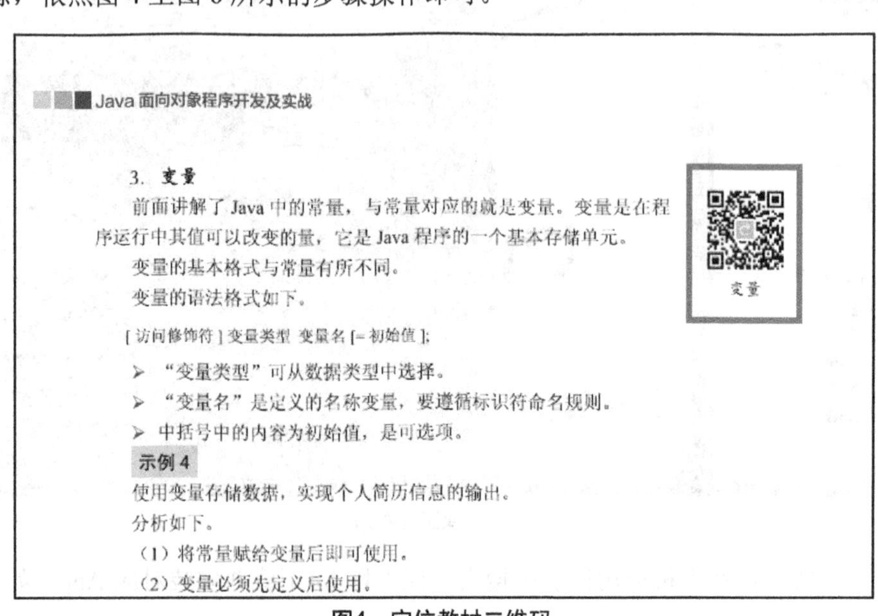
Java 面向对象程序开发及实战

3. **变量**

前面讲解了 Java 中的常量，与常量对应的就是变量。变量是在程序运行中其值可以改变的量，它是 Java 程序的一个基本存储单元。

变量的基本格式与常量有所不同。

变量的语法格式如下。

[访问修饰符] 变量类型 变量名 [= 初始值];

- “变量类型”可从数据类型中选择。
- “变量名”是定义的名称变量，要遵循标识符命名规则。
- 中括号中的内容为初始值，是可选项。

示例 4

使用变量存储数据，实现个人简历信息的输出。

分析如下。

（1）将常量赋给变量后即可使用。

（2）变量必须先定义后使用。

图4　定位教材二维码

图5　使用课工场App“扫一扫”扫描二维码

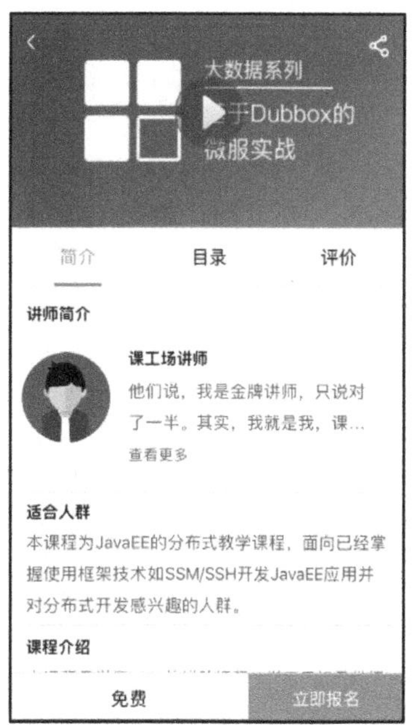

图6　使用课工场App免费观看教材配套视频

3．获取专属的定制化扩展资源

（1）普通读者请访问 http://www.ekgc.cn/bbs 的“教材专区”版块，获取教材所需开发工具、教材中示例素材及代码、上机练习素材及源码、作业素材及参考答案、项目素材及参考答案等资源（注：图 7 所示网站会根据需求有所改版，下图仅供参考）。

图7　从社区获取教材资源

（2）高校老师请添加高校服务 QQ：1934786863（如图 8 所示），获取教材所需开发工具、教材中示例素材及代码、上机练习素材及源码、作业素材及参考答案、项目素材及参考答案、教材配套及扩展 PPT、PPT 配套素材及代码、教材配套线上视频等资源。

图8　高校服务QQ

目　录

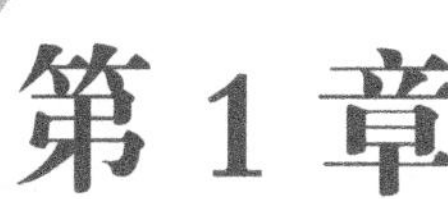

第1章

Hive入门

技能目标

- ➢ 了解Hive的产生背景及部署环境。
- ➢ 了解Hive的架构设计。
- ➢ 掌握Hive的数据存储模型。
- ➢ 能够搭建Hive平台。

本章任务

任务1　了解Hive基础。
任务2　掌握Hive数据存储模型。
任务3　安装配置Hive环境。

本章资源下载

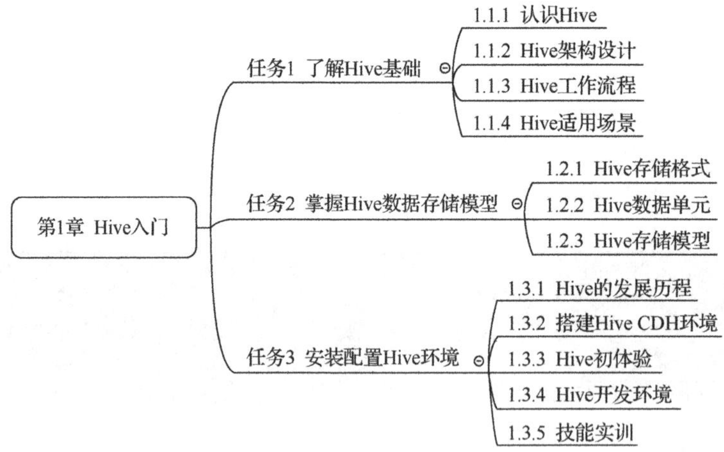

Hadoop 的诞生为大数据解决方案提供了强有力的技术保障。随着 Hadoop 日益完善，Hive 成为了 Hadoop 生态系统中必不可少的工具之一。使用 Hive 可以大大降低将传统数据仓库应用程序转移至 Hadoop 系统上的难度，任何熟悉 SQL 语言的开发者都可以快速地掌握 Hive，并迁移基于 SQL 的应用程序至 Hadoop 系统。

本章将介绍 Hive 的核心构成、工作原理及流程，同时介绍如何搭建 Hive 平台。在本章的学习过程中，读者应特别关注 Hive 与 Hadoop、Hive 与传统关系型数据库之间的关系。

任务 1 了解 Hive 基础

【任务描述】

了解 Hive 产生的背景，Hive 与 Hadoop 以及传统关系型数据库的对比，掌握 Hive 的架构设计与工作原理。

【关键步骤】

（1）了解 Hive 在 Hadoop 生态系统中的位置。

（2）了解 Hive 的产生背景。

（3）了解 Hive 的架构设计。

1.1.1 认识 Hive

1. Hive 产生的背景

Hadoop 中的 MapReduce 计算模型能将计算任务切分成多个小单元，然后分布到各个节点上去执行，从而降低计算成本并提高扩展性。但是使用 MapReduce 进行数据处理的门槛比较高，传统的数据库开发、管理和运维的人员必须掌握 Java 面向 MapReduce API 编程并具备一定的编程基础后，才能使用 MapReduce 处理数据。

然而，Hadoop 分布式文件系统（Hadoop Distributed File System，HDFS）中最关键的一点就是，数据存储在 HDFS 上是没有 Schema（模式）概念的。这里的 Schema 相当于表里面的列、字段、字段名称、字段与字段之间的分隔符等，也可称为 Schema 信息。

在 HDFS 上的数据文件通常是纯文本文件。

那么能否让用户将数据文件从一个现有的数据架构转移到 Hadoop 上来呢？假设该数据架构是基于传统关系型数据库和 SQL 查询的。其实对于大量的 SQL 用户来说，这个问题很难解决。针对这个挑战，Hive 在 Facebook 诞生了。

2. 什么是 Hive

Apache Hive（以下简称 Hive）是一个由 Apache 软件基金会维护的开源项目，由 Facebook 贡献。其前身是 Apache Hadoop 中的一个子项目，现已成为 Apache 顶级项目。

Hive 是一个基于 Hadoop 的数据仓库工具，可以将结构化的数据文件映射为一张数据库表，并提供 SQL 查询功能，同时可以将 SQL 语句转化为 MapReduce 作业进行运行。Hive 具有一系列功能，可以进行数据提取、转化和加载，是一种可以查询和分析存储在 Hadoop 中的大规模数据的工具。总之，Hive 被设计成能够非常方便地进行大数据的汇总、即席查询（ad-hoc）与分析的工具。

对存储在 HDFS 中的数据进行分析与管理时，无须使用 MapReduce 编程而可以通过一个工具来完成相应的操作，这个工具就是 Hive。从这一点上来说，Hive 也是为消除 MapReduce 样板式的编程而产生的。

在 Hadoop 大行其道之前，大部分数据仓库应用程序都须基于关系型数据库实现。数据仓库应用程序通常是建立在数据仓库上的数据应用，包括报表展示、即席查询、数据分析、数据挖掘等。值得注意的是，数据仓库与数据库有所不同，数据仓库源自数据库而又不同于数据库。它们的主要区别是数据仓库适合联机分析处理（On-Line Analytical Processing，OLAP），通常是对某些主题的历史数据进行分析；而数据库适合联机事务处理（On-Line Transaction Processing，OLTP），通常是在数据库联机时对业务数据进行添加、删除、修改、查询等操作。Hive 被设计成数据仓库，其早期版本或新版本在缺省情况（系统默认状态）下并不支持事务，所以并不适合 OLTP。

3. Hive 在 Hadoop 生态系统的位置

Hive 可以将存储在 HDFS 中的结构化数据文件映射成类似关系型数据库表，并接收类 SQL 语句，将其转化为 MapReduce 程序去执行。所以 Hive 必须依赖 Hadoop 而存在，它在 Hadoop 生态系统中的位置如图 1.1 所示。

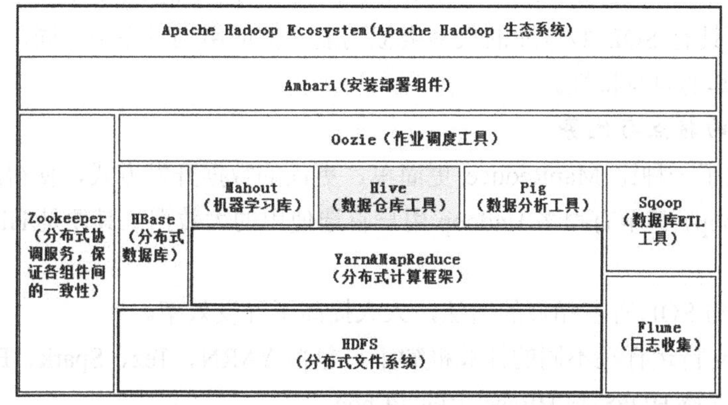

图1.1　Hive在Hadoop生态系统中的位置

从图 1.1 中可以看出，运行 Hive 的必要环境便是 Hadoop 的核心：HDFS、MapReduce 以及 YARN。也可以这样理解，Hive 是由 Hadoop 衍生出来的上层应用之一：SQL on Hadoop。Hive 执行的本质仍然是 MapReduce，但多了一步 SQL 至 MapReduce 的转化操作，所以相同条件下，Hive 在运行时并没有直接编写 MapReduce 执行效率高。

4. Hive 与传统关系型数据库

Hive 与传统关系型数据库（Relational Database Management System，RDBMS）有很多相同的地方，包括查询语言与数据存储模型等。Hive 的 SQL 方言一般被称为 HiveQL，简称 HQL。HQL 并不完全遵循 SQL92 标准，比如 HQL 只支持在 From 子句中使用子查询，并且子查询必须有名字。最重要的是，HQL 须在 Hadoop 上执行，而非传统的数据库。在存储模型方面，数据库、表都是相同的概念，但 Hive 中增加了分区和分桶的概念。

Hive 与 RDBMS 也有其他不同的地方，如在 RDBMS 中，表的 Schema 是在数据加载时就已确定，如果不符合 Schema 则会加载失败；而 Hive 在加载过程中不对数据进行任何验证，只是简单地将数据复制或者移动到表对应的目录下。这也是 Hive 能够支持大规模数据的基础之一。

事务、索引以及更新是 RDBMS 非常重要的特性，鉴于 Hive 的设计初衷，这些特性在开始之初就不是 Hive 设计目标。

更多 Hive 与传统关系型数据库（RDBMS）的对比，如表 1-1 所示。

表 1-1　Hive 与 RDBMS 对比

对比项	Hive	RDBMS
查询语言	HQL	SQL
数据存储	HDFS	块设备、本地文件系统
执行	MapReduce	Executor
执行延迟	高	低
处理数据规模	大	小
事务	0.14 版本后加入	支持
索引	0.8 版本后加入	索引复杂

除了上述区别外，通常 RDBMS 可以用于在线应用中，而 Hive 主要进行离线的大数据分析。Hive 具有 SQL 数据库的很多类似功能，但应用场景完全不同，故在使用的时候要特别注意其的自身特性。

5. Hive 的特点与优势

Hive 提供了一种比 MapReduce 更简单、更优的数据开发方式，使得越来越多的人开始使用 Hadoop，甚至有很多 Hadoop 用户首选使用的大数据工具便是 Hive。Hive 具有以下特点。

- HQL 与 SQL 有着相似的语法，大大提高了开发效率。
- Hive 支持运行在不同的计算框架上，包括 YARN、Tez、Spark、Flink 等。
- Hive 支持 HDFS 与 HBase 上的 ad-hoc。

➢ Hive 支持用户自定义的函数、脚本等。

➢ Hive 支持 Java 数据库连接（Java Database Connectivity，JDBC）与开放数据库连接（Open Database Connectivity，ODBC）驱动，建立了自身与 ETL、BI 工具的通道。

在生产环境中，Hive 具有以下优势。

➢ 可扩展：Hive 可以自由扩展集群的规模，一般情况下无须重启服务。

➢ 可延展：Hive 支持用户自定义函数，用户可根据自己的需求来编写自定义函数。

➢ 可容错：Hive 良好的容错性使得节点出现问题时 SQL 仍可完成执行。

总之，当我们使用 Hive 时，操作接口采用类 SQL 语法，提高了快速开发的能力，避免了编写复杂的 MapReduce 任务，减少了开发人员的学习成本，而且扩展很方便。

1.1.2　Hive 架构设计

1. Hive 架构图

Hive 架构包含 3 个部分。

（1）Hive 客户端（Hive Clients）。Hive 为不同类型应用程序提供不同的驱动，使应用程序可通过 Java、Python 等语言连接 Hive 并进行与 RDBMS 类似的 SQL 查询操作。对于 Java 应用程序，Hive 提供了 JDBC 驱动；对于其他应用程序，Hive 提供了 ODBC 驱动。

（2）Hive 服务端（Hive Services）。客户端必须通过服务端与 Hive 交互，服务端主要包括 CLI、Hive Server、Hive Web Interface、Driver、Metastore 等组件。

（3）Hive 存储与计算（Hive Storage and Computing）。Hive 主要通过元数据存储数据库和 Hadoop 集群进行数据的存储与计算。Hive 的元数据使用 RDBMS 存储，Hive 的数据存储在 HDFS 中，大部分数据查询由 MapReduce 完成。

Hive 的具体架构如图 1.2 所示。

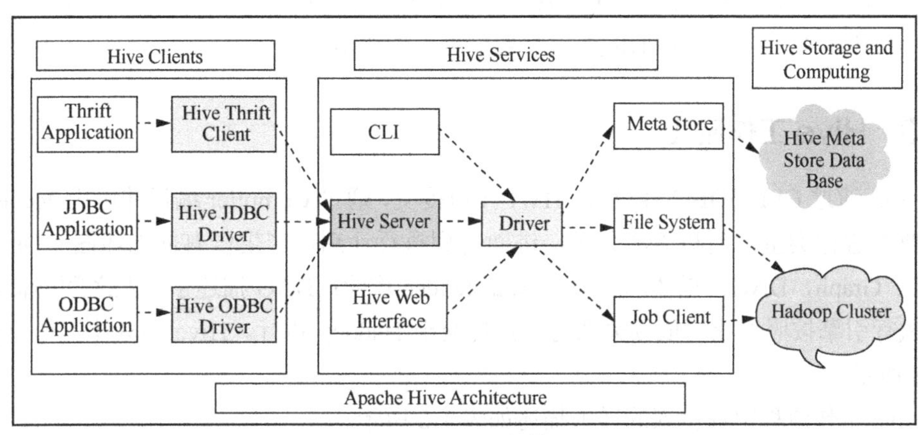

图1.2　Hive架构

2. Hive 服务端组件详解

Hive 服务端主要由以下 3 部分构成。

（1）用户接口

① CLI：控制台命令行方式。CLI 是最基础的连接方式，使用“hive”命令连接。CLI 启动时会同时启动一个 Hive 副本，相当于“hive --service cli”。

② Hive Server：包括 HiveServer1 和 HiveServer2 两种，其中 HiveServer1 在新版中被删除，所以推荐使用 HiverSrver2。HiveServer2 支持一个新的命令行 Shell，其称为 Beeline。Beeline 是一个命令行形式的 JDBC 客户端，用于连接 HiveServer2。Hive 0.11 版中加入了 Beeline，在生产环境中推荐使用 Beeline 连接 Hive。

③ HWI（Hive Web Interface）：通过浏览器访问 Hive。默认端口：9999。

（2）驱动（Driver）组件

该组件包括编译器（Compiler）、优化器（Optimizer）和执行引擎（Executor），它的作用是将 HQL 语句进行解析、编译优化，并生成执行计划，最后调用底层的 MapReduce 计算框架。

（3）元数据服务（Metastore）组件

Hive 中的数据分为两部分，一部分是真实数据，一般存放在 HDFS 中；另一部分是真实数据的元数据，单独存储在关系型数据库中，如 Derby、MySQL 等。元数据用于存储 Hive 中的数据库、表、表模式、目录、分区、索引以及命名空间等信息，是对真实数据的描述。元数据会不断更新变化，所以不适合存储在 HDFS 中。实现任何对 Hive 真实数据的访问均须首先访问元数据。元数据对于 Hive 十分重要，因此 Hive 把 Metastore 服务独立出来，从而解耦 Hive 服务和 Metastore 服务，以保证 Hive 运行的健壮性。

注意

默认情况下，Hive 会使用内置的 Derby 数据库，其只提供有限的单进程存储服务。此时，Hive 不能执行 2 个并发的 Hive CLI 实例，通常被应用于开发、测试环境中。对于生产环境，需要使用 MySQL 或者类似的关系型数据库。

1.1.3 Hive 工作流程

HQL 通过 CLI、JDBC 客户端、HWI 接口提交，通过 Compiler 编译并运用 Metastore 中的数据进行类型检测和语法分析，进而得到执行计划，产生以有向无环图（Directed Acyclic Graph，DAG）描述的一系列 MapReduce 作业；DAG 描述了作业之间的依赖关系，执行引擎按照作业的依赖关系将作业提交至 Hadoop 执行。Hive 的具体工作流程如图 1.3 所示。

Hive 工作流程中各步骤的详细描述如表 1-2 所示。

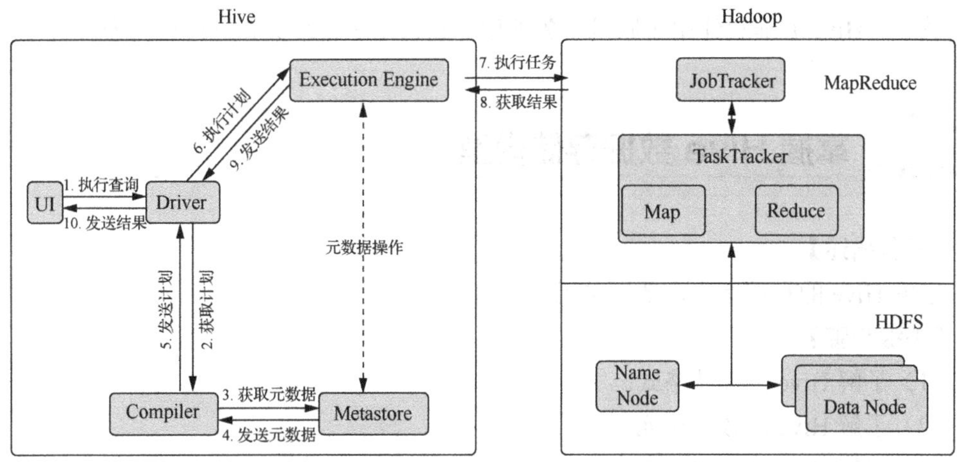

图1.3 Hive工作流程

表 1-2 Hive 工作流程中各步骤的详细描述

序号	操作
1	执行查询：Hive 接口（如命令行或 UI）通过发送查询驱动程序执行查询
2	获取计划：在驱动程序帮助下查询编译器，并分析查询检查语法、查询计划及查询要求
3	获取元数据：编译器发送元数据请求至 Metastore
4	发送元数据：Metastore 发送元数据至编译器以响应之
5	发送计划：编译器检查查询要求，并重新发送查询计划至驱动程序；至此，查询解析与编译完成
6	执行计划：驱动程序发送执行计划至执行引擎
7	执行任务：执行任务的过程是完成一个 MapReduce 工作的过程。执行引擎发送作业至 JobTracker，JobTracker 再把作业分配到 TaskTracker；在 TaskTracker 中查询计划将执行 MapReduce 工作。同时，执行引擎可以通过 Metastore 执行元数据操作
8	获取结果：执行引擎接收来自数据节点的结果
9	发送结果：执行引擎发送结果至驱动程序
10	发送结果：驱动程序发送结果至 Hive 接口

1.1.4 Hive 适用场景

1. **适用场景**

- Hive 适用于非结构化数据的离线分析统计场合。
- Hive 的执行延迟比较高，因此适用于对实时性要求不高的场合。
- Hive 的优势在于处理大数据，因此适用于大数据（而非小数据）处理的场合。

2. **场景技术特点**

- 为超大数据集设计了计算与扩展功能。
- 支持 SQL like 查询语言。
- 支持多表的 join 操作。
- 支持非结构化数据的查询与计算。
- 提供数据存取的编程接口，支持 JDBC、ODBC。

总之，Hive 并非设计用于联机事务处理，建议用于传统数据仓库的开发任务中。

任务 2　掌握 Hive 数据存储模型

【任务描述】

了解 Hive 的存储格式、数据单元与存储模型。

【关键步骤】

（1）了解 Hive 的存储格式。

（2）了解 Hive 的数据单元。

（3）了解 Hive 的存储模型。

1.2.1　Hive 存储格式

Hive 中的数据分为真实数据与元数据，一般来说 Hive 的存储格式是指真实数据的存储格式。

Hive 常用的存储格式包括以下 4 种：TEXTFILE、SEQUENCEFILE、RCFILE 和 ORCFILE。

1. TEXTFILE

常见的 txt、csv、tsv 等文件都属于 TEXTFILE。Hive 默认存储格式为 TEXTFILE，即创建表时如果不单独指定存储格式，则认为存储格式为 TEXTFILE。存储格式为按行存储，内容为普通的文本格式，一般可以使用 cat 命令直接查看。TEXTFILE 格式的数据文件无须任何处理即可导入 Hive，文件每一行为一条记录，记录可用任意分隔符进行列分割，记录之间须以行结束符分割。Hive 将 TEXTFILE 映射为表的过程中，将会逐个字符地判断是否为列分隔符或行分隔符。

TEXTFILE 支持使用 Gzip 压缩，但 Gzip 压缩后的文件将不再支持 MapReduce 分割机制，这意味着压缩后的文件不论有多少个 HDFS 块都只能被一个 Map 任务处理，即失去了使用集群并行处理的优势。

2. SEQUENCEFILE

SEQUENCEFILE 是 Hadoop 提供的一种由“二进制序列化过的 Key/Value 字节流”组成的文本存储文件格式。其由于内容为二进制字节，所以无法通过 cat 命令查看原始字符，这可以抽象地理解为，在 SEQUENCEFILE 中每条记录均以键值对的方式进行存储，仅支持追加。与 TEXTFILE 类似，SEQUENCEFILE 同样按行存储。

Hive 无法直接导入 SEQUENCEFILE 格式的数据文件。数据须首先导入至 TEXTFILE 格式的表中，然后再从 TEXTFILE 格式的表中采用插入方式导入至 SEQUENCEFILE 格式的表。

SEQUENCEFILE 是可分割的文件格式，支持 3 种压缩选项。

（1）NONE：不压缩。

（2）RECORD（默认选项）：记录级压缩，压缩率低。

（3）BLOCK：块级压缩，压缩率高。

3．RCFILE

RCFILE（Record-Columnar File）是 Facebook 开发的一种专门面向列的数据存储格式，不同于 TEXTFILE 和 SEQUENCEFILE，RCFILE 是基于行列混合存储思想的设计。

RCFILE 遵循“先水平划分，再垂直划分”的设计理念：首先把 Hive 表水平切分成多个行组，保证同一行的数据位于同一节点，其次在行组内按照“列”垂直切分，实现列与列的数据在磁盘上呈现为连续的存储块。使用 RCFILE 的优势是既保证了每条记录所有列在同一个 HDFS 块，也可以做到当查询仅针对表中的少数几列时，可跳过不必要的列进行数据读取。RCFILE 存储格式如图 1.4 所示。

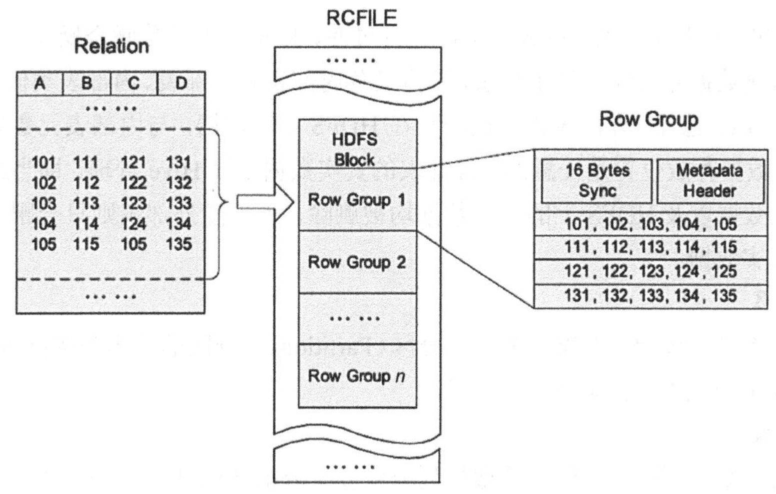

图1.4　RCFILE存储格式

注意

与 SEQUENCEFILE 类似，Hive 在导入 RCFILE 格式的数据文件时也需要经过 TEXTFILE 转换。

RCFILE 是可分割的文件格式，即在每个行组中，元数据头部（Metadata Header）和表格数据段会被分别压缩。对于元数据头部而言，RCFILE 会使用行程长度编码（Run Length Encoding，RLE）算法压缩数据，而对于表格数据段而言，其每个列会通过 Gzip 压缩算法独立压缩。

4．ORCFILE

ORCFILE（Optimized Row-Columnar File）是对 RCFILE 的优化，可以提供一种高效的方法来存储 Hive 数据。ORCFILE 的特点是：支持压缩比很高的压缩算法，文件可切分，提供多种索引，支持复杂的数据结构。

除了上面 4 种常见数据存储格式外，Hive 还支持 Parquet、Avro 等格式，更多 Hive 存储格式的相关内容请扫描二维码获取。

Hive 存储格式

1.2.2 Hive 数据单元

Hive 所有真实数据都存储在 HDFS 中，这样更有利于对数据做分布式计算。为了有效地对真实数据进行管理，根据粒度大小，Hive 将真实数据划分为如下数据单元。

1. **数据库**

数据库（Databases）类似于 RDBMS 中的数据库，在 HDFS 中表现为 hive.metastore.warehouse.dir 目录下的一个文件夹，其本质是用于避免表、视图、分区、列等命名冲突的命名空间。

2. **表**

表（Tables）由列构成，在表上可以进行过滤、映射、连接和联合操作。表在 HDFS 中表现为所属数据库目录下的子目录，具体又分内部表和外部表。内部表类似于 RDBMS 中的表，由 Hive 管理。外部表指向已经存在 HDFS 中的数据，与内部表元数据组织是相同的，但其数据存放位置是任意的。外部表的真实数据不被 Hive 管理，即当删除一张内部表时，元数据以及 HDFS 上的真实数据均被删除，而删除外部表则只会删除元数据而不会删除真实数据。

3. **分区**

每个表都可以按指定的键分为多个分区（Partitions）。分区的作用是提高查询的效率，其在 HDFS 中表现为表目录下的子目录。

4. **分桶**

根据表中某一列的哈希值可将数据划分为多个分桶（Buckets），在 HDFS 中分桶表现为同一个目录下根据哈希散列之后的多个文件。

可以看到，Hive 中的数据单元划分与 RDBMS 的物理模型非常类似。但在 Hive 数据单元划分过程中还必须注意以下几点。

- Hive 表没有主键。
- Hive 表（0.14 版本前）不支持行级操作，新版本中行级操作效率也比较低。
- Hive 表不支持批量 update 操作，但可以先删除、再添加。
- Hive 分区和分桶可以极大提升数据查询效率。

1.2.3 Hive 存储模型

按照数据单元的划分结果，Hive 数据在 HDFS 的典型存储结构中表现为以下形式。

- /数据仓库地址/数据库名称/表名称/数据文件（或分桶数据文件）。
- /数据仓库地址/数据库名称/表名称/分区键/数据文件（或分桶数据文件）。

假设数据仓库地址 hive.metastore.warehouse.dir 为“/hive/warehouse”，则可知如下内容。

（1）"/hive/warehouse /" 表示 Hive 自带的 "default" 数据库位置。

（2）"/hive/warehouse /demo.db" 表示数据仓库中存在 "demo" 数据库。

（3）"/hive/warehouse /demo.db/users" 表示 demo 数据库中存在 users 表。

（4）"/hive/warehouse /demo.db/users/000000_……" 为 demo 中 users 表里的数据文件。

（5）"/hive/warehouse /demo.db/orders/year=2018/part-0000……" 表示 demo 中 orders 表里的数据按照年份进行分区，这里显示的是分区 year=2018 下的数据文件。

Hive 数据存储模型的具体描述如图 1.5 所示。

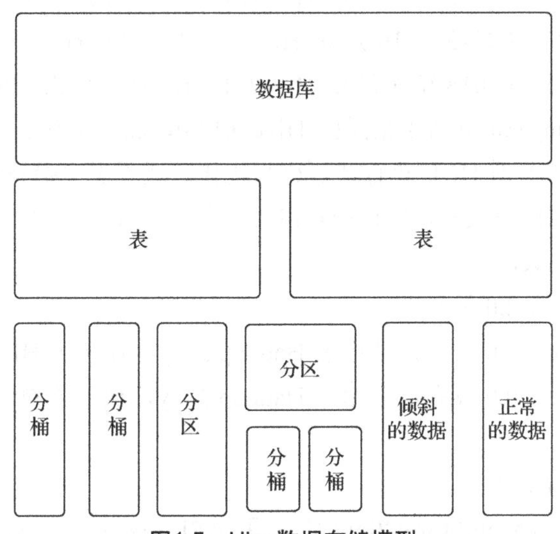

图1.5 Hive数据存储模型

在 Hive 中，数据按分布的均匀情况可以分为正常的数据和倾斜的数据。数据倾斜就是由于数据分布不均匀、数据大量集中到一点上造成的数据热点问题。分区是按照指定的 key 值来划分的，当大量相同的 key 值被分配到一个分区里时就会产生数据倾斜，在后面的章节中还会讨论针对数据倾斜的解决方案。

任务 3 安装配置 Hive 环境

【任务描述】

掌握重要版本的差别，并完成 Hive 环境搭建。

【关键步骤】

（1）下载安装包。

（2）解压安装。

（3）配置属性文件。

（4）启动并验证。

1.3.1 Hive 的发展历程

1. 版本历史

2007 年 8 月由 Facebook 开始开发。

2008 年 8 月开源。

2013 年 2 月 Hortonworks 主导了探针计划（Stinger），这个计划旨在使 Hive 的性能提升 100 倍。Stinger 分为三个阶段完成。

（1）Stinger 阶段一（2013 年 5 月）：Hive 0.11.0，加入了 ORC、HiveServer2 等。

（2）Stinger 阶段二（2013 年 10 月）：Hive 0.12.0，对 ORC 进行了改善。

（3）Stinger 阶段三（2014 年 4 月）：Hive 0.13.0，加入了 Tez 和支持向量化的查询。

2014 年 11 月 Stinger.next 后续阶段：Hive 0.14.0，加入了基于成本的优化器（Cost-based Optimizer，CBO）对 HQL 执行计划进行优化。适用于 Hadoop 1.x.y，2.x.y。

2015 年 2 月 Hive 1.0.0 发布，与 Hive 0.14.0 变化不大。该版本正式移除了 HiveServer1，全面开始使用 HiveServer2。

2015 年 5 月 Hive 1.2.0 发布。

2016 年 2 月 Hive 2.0.0 发布，适用于 Hadoop 2.x.y，加入了 HPLSQL、LLAP 等。

2018 年 5 月 Hive 3.0.0 发布，适用于 Hadoop 3.x.y，加入了物化视图等新功能。

2. 里程碑版本

Hive 0.14.0 引入 CBO。

Hive 1.0.x 为 HiveMetaStoreClient 定义了 API，移除了 HiveServer1，全面使用 HiveServer2。

Hive 2.x 版本中，推出了 140 余项优化与改进。下面简单介绍其四大特性。

（1）开启了 LLAP

Hive 2.1 推出的 LLAP 是下一代分布式计算架构，它能够智能地将数据缓存到多台机器的内存中，并允许所有客户端共享这些缓存的数据，同时还保留了弹性伸缩的能力。通过 LLAP（Live Long and Process），Hive 2.1 进行了极大的性能优化。在开启 LLAP 的 Hive 2.x 与 Hive 1.x 进行对比测试，结果显示 Hive 2.x 的性能提升了约 25 倍。

（2）支持使用 HPL/SQL 的存储过程语言

Hive 2.0.0 推出的 Hive Hybrid Procedural SQL On Hadoop（HPL/SQL）是一个在 Hive 上执行存储过程的 SQL 工具，它可以表达复杂的业务规则。

（3）持续优化成本优化器 CBO

Hive 2.0.0 开始持续不断地优化成本优化器 CBO，尤其是在 BI 业务关注的 TPC-DS 查询上。

（4）提供全面详尽的监控和诊断工具

可以通过新的 HiveServer2 Web UI、LLAP Web UI 和 Tez Web UI 查看 Hive 相关的 HQL 查询以及关联的作业状态和日志，丰富了 Hive 用户的运维和排错手段。

1.3.2　搭建 Hive CDH 环境

1．安装包下载

Hive 可以安装在与 Hadoop 集群连通的任意主机中。本书采用的是 CDH 版本的 hive-1.1.0-cdh5.14.2，读者可以自行在 Cloudera 下载。安装环境要求如下。

- 操作系统：CentOS 7.5
- Java 环境：Java 8
- Hadoop 环境：hadoop-2.6.0-cdh5.14.2
- 数据库：MySQL 5.6

2．解压安装

（1）下载安装包并解压。

```
$ tar -zxvf hive-1.1.0-cdh5.14.2.tar.gz
```

（2）将解压后的文件复制到“/opt”目录下。

```
$ sudo mv hive-1.1.0-cdh5.14.2 /opt/hive-1.1.0-cdh5.14.2
```

注意

如果使用的是“普通用户”进行安装，须加 sudo 才能操作“/opt”目录。

3．Hive CDH 环境搭建

（1）将 Hive 的安装目录添加到环境变量中，命令如下。

```
$ vi ~/.bashrc
export HIVE_HOME= /opt/hive-1.1.0-cdh5.14.2
export PATH= $PATH:$HIVE_HOME/bin
```

添加完成以后，执行如下命令，使环境变量生效。

```
$ source ~/.bashrc
```

（2）在 Hive 配置文件目录$HIVE_HOME/conf 下创建 hive-site.xml，并添加以下代码。

```
<configuration>
    <property>
        <name>hive.metastore.warehouse.dir</name>
        <value>/home/hadoop/hive/warehouse</value>
    </property>
    <!--mysql 数据配置-->
    <property>
        <name>javax.jdo.option.ConnectionURL</name>
        <value>jdbc:mysql://hadoop:3306/hive?createDatabaseIfNotExist=true</value>
    </property>
    <property>
        <name>javax.jdo.option.ConnectionDriverName</name>
        <value>com.mysql.jdbc.Driver</value>
    </property>
```

```
    <!-- Hive 连接 MySQL 的用户名和密码-->
    <property>
        <name>javax.jdo.option.ConnectionUserName</name>
        <value>hive</value>
    </property>
    <property>
        <name>javax.jdo.option.ConnectionPassword</name>
        <value>hive</value>
    </property>
    <!--配置 Hive 临时文件存储地址-->
    <property>
       <name>hive.exec.scratchdir</name>
        <value>/home/hadoop/hive/data/hive-${user.name}</value>
        <description>Scratch space for Hive jobs</description>
    </property>
    <property>
        <name>hive.exec.local.scratchdir</name>
        <value>/home/hadoop/hive/data/${user.name}</value>
        <description>Local scratch space for Hive jobs</description>
    </property>
</configuration>
```

（3）复制 MySQL JDBC 驱动 jar 文件到“$HIVE_HOME/lib”目录下，同时修改目录的可写权限。

```
$ cp mysql-connector-java-5.1.44-bin.jar $HIVE_HOME/lib
# 验证
$ ll $HIVE_HOME/lib|grep mysql
-rw-r--r--   1 hadoop hadoop     999635 8 月   29 2017 mysql-connector-java-5.1.44-bin.jar
```

（4）修改 Hive 环境脚本 hive-env.sh 的内容。

```
# 设置 HADOOP_HOME 指定 Hadoop 安装目录
HADOOP_HOME=/opt/hadoop-2.6.0-cdh5.14.2
# 设置 Hive 配置目录
export HIVE_CONF_DIR=/opt/hive-1.1.0-cdh5.14.2/conf
# 设置扩展类路径
export HIVE_AUX_JARS_PATH=/opt/hive-1.1.0-cdh5.14.2/lib
```

（5）配置日志。

```
#创建日志输出目录
$ mkdir /opt/hive-1.1.0-cdh5.14.2/logs
#创建 hive-log4j.properties
$ mv $HIVE_HOME/conf/hive-log4j.properties.template $HIVE_HOME/conf/hive-log4j.properties
#修改 hive-log4j.properties
$ vi $HIVE_HOME/conf/hive-log4j.properties
#指定日志输出目录
hive.log.dir=/opt/hive-1.1.0-cdh5.14.2/logs
```

（6）启动 Hive，显示所有数据库。

```
$ cd $HIVE_HOME/bin
$ hive
hive> show databases;
OK
default
Time taken: 8.651 seconds, Fetched: 1 row(s)
```

注意

启动 Hive 前要先启动 MySQL，并在 MySQL 中添加 Hive 用户名和密码。参考步骤如下。

```
create user 'hive'@'%' identified by 'hive';
grant all on *.* to 'hive'@'hadoop' identified by 'hive';
set password for hive@hadoop=password('hive');
flush privileges;
```

至此，Hive 安装已完成。Hive CDH 环境搭建视频请扫描二维码查看。

Hive CDH 环境搭建

1.3.3 Hive 初体验

WordCount 词频统计是 MapReduce 的经典案例，其可对文本文件中某单词出现的次数进行统计。WordCount 的官方源码如下。

```
package org.apache.hadoop.examples;
import java.io.IOException;
import java.util.StringTokenizer;
import org.apache.hadoop.conf.Configuration;
import org.apache.hadoop.fs.Path;
import org.apache.hadoop.io.IntWritable;
import org.apache.hadoop.io.Text;
import org.apache.hadoop.mapreduce.Job;
import org.apache.hadoop.mapreduce.Mapper;
import org.apache.hadoop.mapreduce.Reducer;
import org.apache.hadoop.mapreduce.lib.input.FileInputFormat;
import org.apache.hadoop.mapreduce.lib.output.FileOutputFormat;
import org.apache.hadoop.util.GenericOptionsParser;
public class WordCount {
  public static class TokenizerMapper
       extends Mapper<Object, Text, Text, IntWritable>{
    private final static IntWritable one = new IntWritable(1);//VALUEOUT
    private Text word = new Text();//KEYOUT
    public void map(Object key, Text value, Context context
                    ) throws IOException, InterruptedException {
      StringTokenizer itr = new StringTokenizer(value.toString());
      while (itr.hasMoreTokens()) {
```

```
            word.set(itr.nextToken());
            context.write(word, one);
        }
      }
    }
    public static class IntSumReducer
          extends Reducer<Text,IntWritable,Text,IntWritable> {
      private IntWritable result = new IntWritable();
      public void reduce(Text key, Iterable<IntWritable> values,
                         Context context
                         ) throws IOException, InterruptedException {
        int sum = 0;
        for (IntWritable val : values) {
          sum += val.get();
        }
        result.set(sum);
        context.write(key, result);
      }
    }
    public static void main(String[] args) throws Exception {
      Configuration conf = new Configuration();
      String[] otherArgs = new GenericOptionsParser(conf, args).getRemainingArgs();
      if (otherArgs.length < 2) {
        System.err.println("Usage: wordcount <in> [<in>...] <out>");
        System.exit(2);
      }
      Job job = Job.getInstance(conf, "word count");
      job.setJarByClass(WordCount.class);
      job.setMapperClass(TokenizerMapper.class);
      job.setCombinerClass(IntSumReducer.class);
      job.setReducerClass(IntSumReducer.class);
      job.setOutputKeyClass(Text.class);
      job.setOutputValueClass(IntWritable.class);
      for (int i = 0; i < otherArgs.length - 1; ++i) {
        FileInputFormat.addInputPath(job, new Path(otherArgs[i]));
      }
      FileOutputFormat.setOutputPath(job,
        new Path(otherArgs[otherArgs.length - 1]));
      System.exit(job.waitForCompletion(true) ? 0 : 1);
    }
  }
```

基于上述源码可知，即便是很简单的 MapReduce 程序也含有 60 多行代码。那如果使用 Hive 来完成 WordCount 需要多少行代码呢？

（1）准备数据文件 mywords.txt。

```
$ vi mywords.txt
# 输入如下单词，行内以空格分隔，行间换行符分隔
Hello World
Hello Hive
Hello Hadoop
# 上传至 HDFS
$ hdfs dfs -mkdir -p /data/wordcount
$ hdfs dfs -put mywords.txt /data/wordcount
```

（2）进入 Hive CLI，编写 HQL 语句以完成 WordCount。

```
$ hive
# 创建表
hive> create external table lines(ling string);
# 装载数据
hive> load data inpath '/data/wordcount' overwrite into table lines;
# 查询统计
hive> select word,count(*) as wc from lines lateral view explode(split(line,' ')) t1 as word group by word;
```

（3）执行结果如下。

```
hive> create external table lines(line string);
OK
Time taken: 4.742 seconds
hive> load data inpath '/data/wordcount' overwrite into table lines;
Loading data to table default. lines
Table default.lines stats: [numFiles=1, totalSize=36]
OK
Time taken: 0.514 seconds
hive> select word,count(*) as wc from lines lateral view explode(split(line,' ')) tl as word group by word;
Query ID = root_20181129142424_2948e218-ddd7-4e7c-803e-bcd3d21db21f
Total jobs = 1
Launching Job 1 out of 1
Number of reduce tasks not specified. Estimated from input data size: 1
In order to change the average load for a reducer (in bytes):
  set hive.exec.reducers.bytes.per.reducer=<number>
In order to limit the maximum number of reducers:
  set hive.exec.reducers.max=<number>
Hadoop job information for Stage-1: number of mappers:
14:24:46,642   Stage-1 map = 0%,reduce = 0%
14:24:55,246   Stage-1 map = 100%,reduce = 0%,Cumulative CPU 4.41 sec
14:25:04,715   Stage-1 map = 100%, reduce = 100%,Cumulative CPU 9.13 sec
MapReduce Total cumulative CPU time: 9 seconds 130 msec
Ended Job = job_1543310100051_0033
MapReduce Jobs Launched:
```

```
Stage-Stage-1: Map: 1 Reduce: 1 Cumulative CPU: 9.13 sec HDFS Read: 9637 HDFS Write: 32 SUCCESS
Total MapReduce CPU Time Spent: 9 seconds 130 msec
OK
Hadoop 1
Hello 3
Hive 1
World 1
Time taken: 47.78 seconds, Fetched: 4 row(s)
hive>
```

由上述代码可知，仅用 3 条 HQL 语句即可完成 WordCount，相比 MapReduce 的开发效率有了极大的提升。但是，WordCount 源码和适用 Hive 完成的 WordCount 代码哪个执行效率更高呢？

前面提到，HQL 最终会被翻译成 MapReduce 去执行，针对结果而言两者的本质没有区别，但 HQL 翻译过程需要额外的开销，所以 WordCount 源码执行效率更高，但是站在开发效率的角度考虑，损失 Hive 的这点性能是完全值得的。

1.3.4 Hive 开发环境

在开发过程中，Hive 提供了两种命令行工具：CLI 与 Beeline，分别对应 hive 与 beeline 命令。CLI 工具一般在 Hive 主机上使用，而 Beeline 支持远程连接。使用 Beeline 需要首先在 Hive 主机中启动 HiveServer2 服务。

使用 Beeline 连接 Hive：

```
# 在 Hive 主机执行
$ hive --service hiveserver2&
# 可在任意客户端执行
$ beeline -u jdbc:hive2://localhost:10000
# 交互式执行 HQL
0:jdbc:hive2://localhost:10000> show tables;
```

其中，10000 为 HiveServer2 默认端口，其可通过“hive.server2.thrift.port”参数进行修改。默认情况下启动 HiveServer2 会同时启动一个内置的元数据服务。

注意

HiveServer2 进程建议在后台运行：

```
nohup hive --service hiveserver2&
```

CLI 和 Beeline 均支持两种模式：命令行模式和交互模式。前面介绍的方式均属于交互模式。Hive 支持使用命令直接运行 HQL 语句或者包含一系列 HQL 语句的文件。

【语法】

```
hive -e <HQL>
hive -f <HQL FILE>
```

```
beeline –u <JDBC Url> -e <HQL>
beeline –u <JDBC Url> -f <HQL FILE>
```

下面分别使用 CLI 和 Beeline 查看 Hive 所有数据库：

```
$ hive -e 'show databases;'
$ beeline -u jdbc:hive2://localhost:10000 -e 'show databases;show tables'
```

注意

请使用引号包含 HQL 语句，多条 HQL 之间使用分号分隔。

Hive 也可以使用文件来管理 HQL，然后通过“-f”选项执行。

```
$ vi showdbs.hql
# 输入 HQL 语句
show databases;
$ hive -f showdbs.hql
$ beeline -u jdbc:hive2://localhost:10000 -f showdbs.hql
```

除了 CLI 和 Beeline，还有多种第三方工具均可以很方便地进行 Hive 开发，下面分别对它们进行简单介绍。

1. HUE

Hadoop 用户经验（Hadoop User Experience，HUE）集成在 CDH 中，是一个开源 SQL 分析平台，有时也称为 Hadoop UI。通过 HUE 可以在浏览器中与 Hadoop 集群中的多种组件（如 Hive、Pig、Impala 等）进行交互以实现数据的分析处理。

如果选用了 Cloudera 的 CDH 版本，则适合选择 HUE 进行 Hive 开发，开发界面如图 1.6 所示。

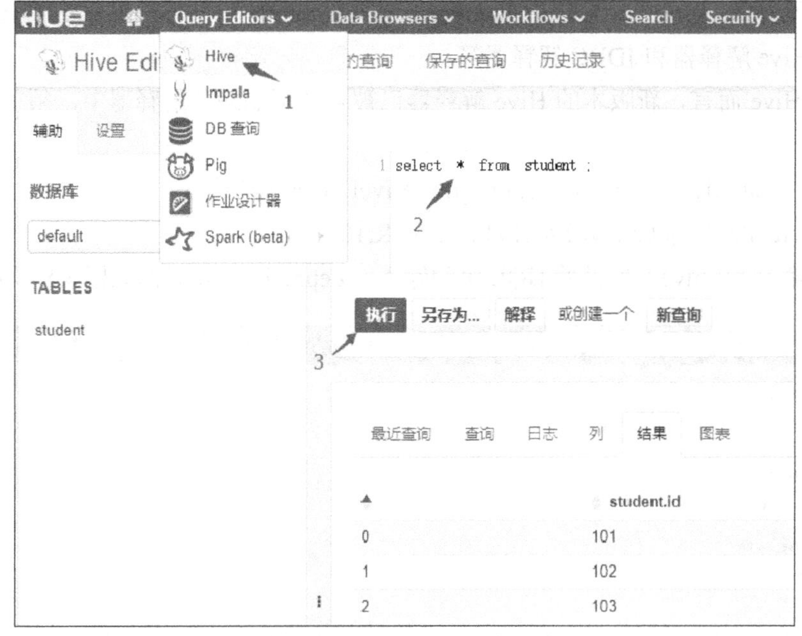

图1.6　选择HUE开发Hive界面

2. Ambari Hive View

Apache Ambari 是一种基于 Web 的工具，支持 Apache Hadoop 集群的管理和监控。Hive View 是 Ambari 中的一个组件。Hortonworks 的 HDP 集成了 Ambari，所以使用 HDP 平台时选择 Ambari 更合适。Ambari 开发界面如图 1.7 所示。

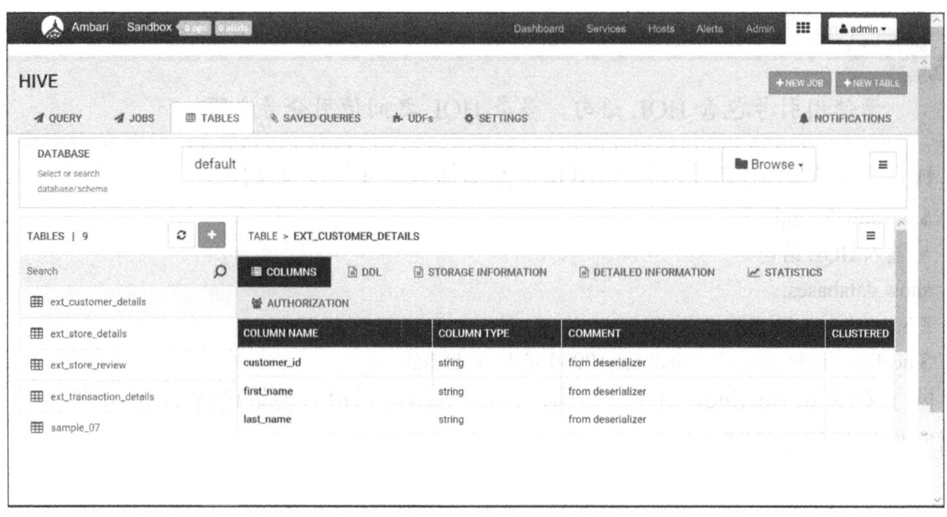

图1.7 Ambari Hive View开发界面

3. Zeppelin

Apache Zeppelin 是一个基于 Web 的 Notebook，其支持多种语言、多种环境下的交互式数据分析，包括：Python（Spark）、Scala（Spark）、Spark SQL、Hive、Shell 等。

Zeppelin 核心概念是解释器（Interpreter）插件。Interpreter 允许用户指定一种语言或者数据处理引擎。当前已经实现的 Interpreter 有 Python 解释器、Spark SQL 解释器、Shell 解释器、Hive 解释器和 JDBC 解释器等。

针对 Hive 而言，新版本的 Hive 解释器已被合并至 JDBC 解释器中，须首先做如下配置。

- default.driver：org.apache.hive.jdbc.HiveDriver（默认）。
- default.url：jdbc:hive2://HIVESERVER:10000。

然后输入“%hive”后开始 HQL 语句输入。Zeppelin 开发界面如图 1.8 所示。

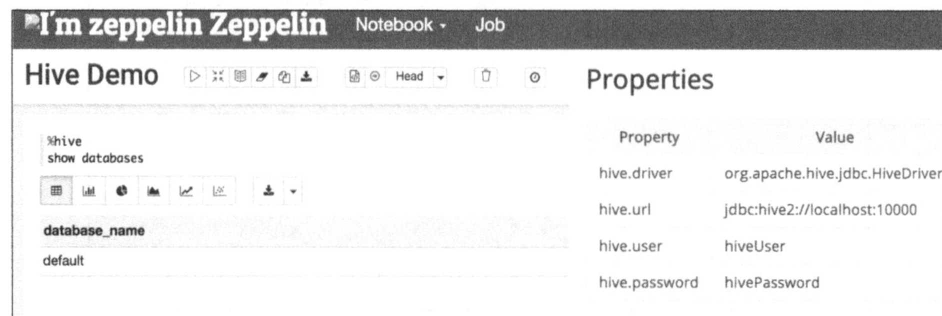

图1.8 Zeppelin开发界面

1.3.5 技能实训

按照任务 3 中的步骤，搭建自己的 Hive CDH 环境，并使用 Hive 完成 WordCount。关键步骤如下。

（1）下载安装包。

（2）配置 Hive 相关环境变量。

（3）修改 hive-site.xml 和 hive-env.sh 文件。

（4）添加 Hive 连接 MySQL 的用户名、密码。

（5）启动 MySQL。

（6）启动 Hive。

（7）完成 WordCount。

本章小结

- Hive 是 Hadoop 上处理结构化数据的数据仓库基础工具，它用来处理存储在 Hadoop 上的海量数据。使用 Hive 可使数据查询和分析变得更简单。
- Hive 提供了 HQL 以完成 MapReduce 开发，HQL 使得传统数据库开发人员更容易使用 Hadoop。
- Hive 依赖于 Hadoop 的 HDFS 与 YARN，安装 Hive 前应先安装 Hadoop。
- Hive 与 RDBMS 在使用方法上类似，但应用场景有着显著区别。
- Hive 架构主要包括：CLI、HiveServer2、HWI、Driver、Metastore。
- Hive 数据存储模型与 RDBMS 的物理模型类似，分区与分桶是 Hive 为了提升查询性能而提出的概念。
- Hive 的元数据是对真实数据的描述，通常单独存储在 MySQL 中，并且元数据服务与 Hive 服务可以相互隔离。
- CLI 与 Beeline 是 Hive 基本操作、管理、开发工具，其中 Beeline 应用更广泛。二者均有两种运行模式：交互模式与命令行模式。
- HUE、Ambari 及 Zeppelin 等工具均支持 Hive 开发。

本章作业

一、简答题

1．简述 Hive 架构的组成。

2．Hive 的数据单元有哪些？它们的作用分别是什么？

3．CLI 与 Beeline 的区别是什么？

二、编码题

1．现有 A、B 两张数据表，格式如表 1-3 和表 1-4 所示。

表 1-3　A 数据表

user_id	type_id
1001	2
1002	3
1003	1
1004	2
1005	2
……	……

表 1-4　B 数据表

user_id
1001
1002
1008
1009
1010
……

使用 SQL 语句实现查询：在 A 数据表中有而 B 数据表中没有的 user_id。

2．库存表 store 中有类型为铅笔的物品（prod=3），出库类型（type=1），入库类型（type=2），num 为数量，具体数据如表 1-5 所示。请用 SQL 求出铅笔的库存量。

表 1-5　store 表

prod	type	num
3	1	5
3	1	6
3	2	11
3	2	10
……	……	……

3．存在一张通话记录表，其字段包括：手机号（mdn），通话状态（type：主叫 1，被叫 2），通话省份编号（pro），具体数据如表 1-6 所示。请用 SQL 抽取每个省份通话次数最多的 100 个手机号。

表 1-6　通话记录表

mdn	type	pro
13301034567	1	5
13301034567	2	6
13301034567	1	7
13301034567	2	8
……	……	……

第 2 章

Hive 数据库及表操作

技能目标

- 掌握 Hive 的数据类型。
- 掌握 Hive 数据库和表操作。
- 理解 Hive 内部表和外部表。
- 理解 Hive 数据分区和分桶。
- 掌握 Hive Shell 命令行模式。
- 掌握 Hive Java API 编程。

本章任务

任务 1　熟悉 Hive 数据类型。

任务 2　使用 Hive 管理雇员信息。

任务 3　使用 Hive Java API 操作雇员表。

本章资源下载

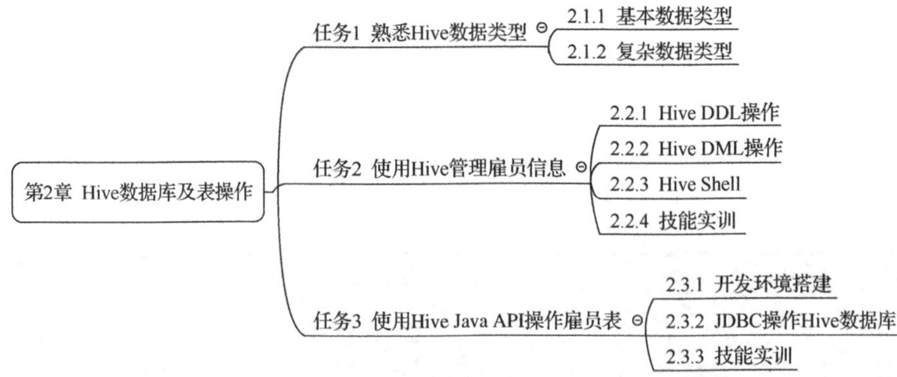

在 Hive 中，同样提供了与 DDL、DML 和 DQL 类似的操作。本章将通过介绍 Hive DDL、DML 和查询来讲解如何对 Hive 数据库和表进行操作。

任务 1 熟悉 Hive 数据类型

【任务描述】

了解 Hive 常用的基本数据类型和复杂数据类型。

【关键步骤】

（1）了解 Hive 的基本数据类型。

（2）了解 Hive 的复杂数据类型。

2.1.1 基本数据类型

Hive 中的基本数据类型也称为原始类型，包括整数、小数、文本、布尔、二进制以及时间类型。

整数：TINYINT、SMALLINT、INT、BIGINT

小数：FLOAT、DOUBLE、DECIMAL

文本：STRING、CHAR、VARCHAR

布尔：BOOLEAN

二进制：BINARY

时间：DATE、TIMESTAMP、INTERVAL

上述数据类型都是对 Java 中的接口的实现，所以类型的具体行为细节和 Java 中对应的类型完全一致。比如 STRING 类型实现的是 Java 中的 String，FLOAT 类型实现的是 Java 中的 float。

注意

- HiveQL 对大小写不敏感。
- Hive 中所有类型均为保留字。

1. **整数类型**

Hive 具有 4 种带符号的整数类型：TINYINT、SMALLINT、INT、BIGINT，分别对应 Java 中的 byte、short、int、long，字节长度分别为 1、2、4、8 字节。在使用整数变量时，默认情况下为 INT，如果要声明为其他类型，须通过后缀来标识，如表 2-1 所示。

表 2-1 整数类型 INT 声明为其他整数类型时的后缀标识

类型	后缀	例子
TINYINT	Y	100Y
SMALLINT	S	100S
BIGINT	L	100L

2. **小数类型**

小数类型也称为浮点类型，包括 FLOAT 和 DOUBLE 两种，对应 Java 中的 float 和 double，分别为 32 位和 64 位浮点数。

此外，Hive 还可以使用 DECIMAL 来表示任意精度的小数，类似于 Java 中的 BigDecimal，其通常在货币当中使用。DECIMAL 语法和示例如下。

【语法】

DECIMAL(precision[,scale])

其中，precision 表示固定精度，最大为 38；scale 表示小数位数。

例如：

- DECIMAL(5,2)存储-999.99 到 999.99 的数字；
- DECIMAL(5)表示-99999 到 99999 的数字；
- DECIMAL 等同于 DECIMAL(10,0)。

3. **文本类型**

文本类型也称为字符串类型，使用单引号或双引号来指定。Hive 有 3 种类型用于存储文本。

（1）STRING 存储可变长的文本，对长度没有限制。理论上 STRING 的存储空间为 2GB，但是存储特别大的对象时效率会受影响，此时可以考虑使用 Sqoop 提供的大对象支持。

（2）VARCHAR 与 STRING 类似，但其存储字符串的长度要求在 1～65355 之间，超出部分将被截断，例如 VARCHAR(20)。

（3）CHAR 以固定长度来存储字符串，最大固定长度为 255，例如 CHAR(20)。

注意

VARCHAR(20)与 CHAR(20)的区别是：如果存入的字符串长度为 10，则 VARCHAR(20)占用实际字符串长度为 10，而 CHAR(20)占用实际字符串长度仍为 20，未使用的空间将会用空格填充。

4. **布尔及二进制类型**

布尔（BOOLEAN）类型值有 true 和 false 两种。

二进制类型 BINARY 用于存储变长的二进制数据，如 10110。

5. **时间类型**

DATE 类型用于描述特定的年月日，以 yyyy-MM-dd 格式表示，例如 2018-12-12。DATE 类型不包含时间，所表示的日期范围为 0000-01-01 至 9999-12-31。

TIMESTAMP 类型存储纳秒级别的时间戳。TIMESTAMP 与时区无关，存储 UNIX 纪元（1970-01-01 00:00:00）的偏移量，其常见用法如下。

（1）unix_timestamp()函数

- 使用 unix_timestamp()函数得到当前时间戳。
- 使用 unix_timestamp(string date)将“yyyy-MM-dd HH:mm:ss”格式的字符串转为时间戳。
- 使用 unix_timestamp(string date,string format)将指定字符串（如“2018-12-12”）按指定格式（如“yyyy-MM-dd”）转为时间戳。

（2）from_unixtime()函数

使用 from_unixtime(timstamp,' yyyy-MM-dd HH:mm:ss ')将指定时间戳以指定格式显示。

（3）cast()函数

cast()函数用于在 TIMESTAMP、DATE、STRING 类型之间做转换，举例如下。

- cast(date as date)：返回相同的日期。
- cast(date as timestamp)：基于本地时区，返回对应 00:00:00 时间戳的日期。
- cast(date as string)：日期被转换为“yyyy-MM-dd”格式的字符串。
- cast(timestamp as date)：基于本地时区确定时间戳的年月日，并将其作为值返回。
- cast(string as date)：如果字符串的格式为“yyyy-MM-dd”，则返回对应的日期，否则返回 NULL。

INTERVAL 类型表示时间间隔，其仅被 Hive 1.2.0 之后的版本支持。

2.1.2 复杂数据类型

在 SQL 的表设计中，字段通常不能被再分解，这意味着每一个字段不能再被分割成多个字段。如表 2-2 所示订单不满足第一范式的要求，因此我们将无法方便地使用 SQL 对订单 1001 中的商品列表进行添加或删除操作。

表 2-2 订单

订单	商品
1001	["空调","Wi-Fi","西瓜","沙发"]

注意

第一范式是指一个关系模式中的所有属性都是不可分的基本数据项。

与 SQL 不同，HiveQL 没有上述局限性。Hive 的表字段不仅可以是基础数据类型，还可以是复杂数据类型（Complex Types）。Hive 有 4 种常用的复杂数据类型，分别是数组（ARRAY）、映射（MAP）、结构体（STRUCT）和联合体（UNIONTYPE）。

1. ARRAY 和 MAP

ARRAY 是具有相同类型变量的集合。这些变量称为数组的元素，每个数组元素都有一个索引编号，编号从零开始。

数据格式：['Apple', 'Orange', 'Mongo']

定义示例：array<string>

使用示例：a[0]= 'Apple'

其中，a 表示表字段名，该字段的类型为 ARRAY，允许存储的数据为 string，起始访问索引为 0。

MAP 是一组键值对集合，key 只能是基本类型，值可以是任意类型。

数据格式：{'A':'Apple', 'O':'Orange'}

定义示例：map<string,string>

使用示例：b['A']='Apple'

其中，b 表示表字段名，该字段的类型为 MAP，允许存储一个<string,string>类型的键值对，直接按键访问即可。

2. STRUCT

STRUCT 封装了一组有名字的字段（Named Field），其类型可以是任意的基本类型，结构体内的元素使用“.”来访问。

数据格式：{'Apple',2}

定义示例：struct<fruit:string,weight:int>

使用示例：c.weight=2

其中，c 表示表字段名，该字段的类型为 STRUCT，包含两个元素“fruit”和“weight”，对元素的访问方式类似于 Java 中访问对象的属性。

3. UNIONTYPE

在给定的任何一个时刻，UNIONTYPE 类型可以保存指定数据类型中的任意一种。其类似于 Java 中的泛型，在任一时刻只有其中的一个类型生效。

定义示例：uniontype<data_type,data_type,…>

使用时，每个 UNIONTYPE 实际对应的类型都会通过一个整数来表示，这个整数位声明时的索引从 0 开始。例如：

```
hive> create table union_test(
    > foo UNIONTYPE<int,double,array<string>,struct<a:int,b:string>>);
```

上面创建了 union_test 表，表中唯一字段 foo 的类型为 UNIONTYPE，在 foo 字段上支持如下文本输入：

```
0^B1
1^B2.0
2^Bthree^Cfour
3^B5^Csix
```

其中“^B”为集合元素分隔符，“^C”为 MAP 键分隔符。关于分隔符用法稍后再详细介绍。

提示

“^B”和“^C”的输入方式：在 vi 界面中先按组合键 Ctrl+V，再按组合键 Ctrl+B 或 Ctrl+C。

第一行表示数据类型为“int”，值为“1”。

第二行表示数据类型为“double”，值为“2.0”。

第三行表示数据类型为“array<string>”，值为“['three','four']”。

第四行表示数据类型为“struct<a:int,b:string>”，值为“{'a':5,'b':'six'}”。

将上述文本另存为文件“data.txt”，再加载至表 union_test 中。

```
hive> load data local inpath '/home/hadoop/data.txt' overwrite into table union_test;
    > select * from union_test;
```

输出结果为：

```
{0:1}
{1:2.0}
{2: ['three','four']}
{3: {'a':5,'b':'six'}}
```

通过上述操作可以发现：UNIONTYPE 的特点是可以存储任何已定义的类型，但是同一时刻只能使用其中一种类型。

注意

目前 Hive 对 UNIONTYPE 类型的支持还不够完善，在这种类型的字段上进行 join、where、group by 操作将会失败。

任务 2 使用 Hive 管理雇员信息

【任务描述】

基于雇员信息文件 employee.txt，完成雇员表的创建、数据加载与查询。

【关键步骤】

（1）Hive 数据表定义操作（DDL）。

（2）Hive 数据加载操作（DML）。

（3）Hive 查询操作。

（4）Hive Shell 操作。

2.2.1　Hive DDL 操作

Hive DDL 用于定义 Hive 数据库模式，其命令包括 create、drop、alter、truncate、show 和 describe 等，主要是对数据库和表进行创建、修改、删除等操作。

1．***数据库操作***

（1）创建数据库

【语法】

```
CREATE (DATABASE|SCHEMA) [IF NOT EXISTS] database_name
[COMMENT database_comment]
[LOCATION hdfs_path]
[WITH DBPROPERTIES   (property_name=property_value,……)];
```

示例 2-1

创建 empdb 数据库。

关键代码：

```
hive> create database empdb;
    > show databases;
    > use empdb;
```

其中，“show databases;” 显示数据库列表，“use empdb;” 选择 empdb 作为当前操作的数据库。

（2）修改数据库

【语法】

```
ALTER (DATABASE|SCHEMA) database_name
SET DBPROPERTIES(property_name=property_value,……)
```

（3）删除数据库

【语法】

```
DROP (DATABASE|SCHEMA) [IF EXISTS] database_name [RESTRICT|CASCADE];
```

默认情况下使用 RESTRICT 删除数据库。如果数据库非空，则使用 RESTRICT 删除数据库将会失败，此时须使用 CASCADE 级联删除数据库。

2．***表操作***

（1）创建表

【语法】

```
CREATE [TEMPORARY] [EXTERNAL] TABLE [IF NOT EXISTS] [db_name.]table_name
  [(col data_type [COMMENT col_comment], ... [constraint_specification])]
  [COMMENT table_comment]
```

```
[PARTITIONED BY (col data_type [COMMENT col_comment], ...)]
[CLUSTERED BY (col, ...) [SORTED BY (col [ASC|DESC], ...)] INTO num_buckets BUCKETS]
[SKEWED BY (col, ...)
    ON ((col_value, col_value, ...), (col_value, col_value, ...), ...)
    [STORED AS DIRECTORIES]
[
  [ROW FORMAT row_format]
  [STORED AS file_format]
    | STORED BY 'storage.handler.class.name' [WITH SERDEPROPERTIES (...)]
]
[LOCATION hdfs_path]
[TBLPROPERTIES (property_name=property_value, ...)]
```

创建表语法中的关键点介绍如下。

- CREATE TABLE：创建一个指定名称的表，如果存在同名表，则抛出异常。可使用“IF NOT EXISTS”忽略该异常。
- EXTERNAL：表示该表为外部表，须同时指定实际数据的 HDFS 路径 LOCATION。注意：当表被删除时，实际数据不会被删除。
- TEMPORARY：指定该表为临时表，被 Hive 0.14 之后的版本支持。临时表只对当前会话有效，会话退出后临时表自动删除。注意，临时表不支持分区与索引。
- PARTITIONED BY：创建表的时候可以同时为表创建一个或多个分区。
- CLUSTERED BY：分桶，让数据能够均匀地分布在表的各个数据文件中。
- ROW FORMAT：用于指定序列化与反序列化器（Serializer/Deserializer，SerDe）。Hive 使用 SerDe 读写表的每一行数据。如果 ROW FORMAT 未指定或者指定了“ROW FORMAT DELIMITED”，Hive 将使用内置的 SerDe 来确定表中具体字段的数据。

【语法】

```
ROW FORMAT
    DELIMITED
        [FIELDS TERMINATED BY char]
        [COLLECTION ITEMS TERMINATED BY char]
        [MAP KEYS TERMINATED BY char]
        [LINES TERMINATED BY char]
    |SERDE serde_name [WITH SERDEPROPERTIES (...)]
```

- STORED AS：如果表中数据是文本数据，则使用 TEXTFILE；如果数据需要压缩，可使用 SEQUENCEFILE 等。

示例 2-2

创建 employee_external 外部表，其中/user/hadoop/employee/employee.txt 雇员信息文件的内容如图 2.1 所示。

```
Michael|Montreal,Toronto|Male,30|DB:80|Product:DeveloperEOTLead
Will|Montreal|Male,35|Perl:85|Product:Lead,Test:Lead
Shelley|New York|Female,27|Python:80|Test:Lead,COE:Architect
Lucy|Vancouver|Female,57|Sales:89,HR:94|Sales:Lead
```

图2.1　雇员信息文件

分析：

➢ 每行数据包括 5 个字段，即姓名、工作地点、性别与年龄、技能值以及部门与职务，各字段以“|”分隔。

➢ 姓名为字符串类型。

➢ 有些雇员的工作地点出现多次，故工作地点应该为 ARRAY 类型。

➢ 性别与年龄适合用 STRUCT 类型。

➢ 每个雇员技能名称不同，所以技能值不适合 STRUCT 类型，而适合 MAP 类型。

➢ 有些雇员在同一部门担任多个职务，故部门与职务数据可用 MAP<STRING, ARRAY<STRING>>存储，其中 KEY 为部门，VALUE 为可能担任的多个职务。图 2.1 中“EOT”是分隔符，对应 ASCII 码'\004'（^D）。

关键代码：

```
CREATE EXTERNAL TABLE IF NOT EXISTS employee_external (
    name string,
    work_place ARRAY<string>,
    sex_age STRUCT<sex:string,age:int>,
    skills_score MAP<string,int>,
    depart_title MAP<STRING,ARRAY<STRING>>
)
ROW FORMAT DELIMITED
  --指定字段分隔符，默认是 ASCII 码'\001'（^A）
    FIELDS TERMINATED BY '|'
  --指定集合元素分隔符，默认是 ASCII 码'\002'（^B）
    COLLECTION ITEMS TERMINATED BY ','
  --指定 MAP 键分隔符，默认是 ASCII 码'\003'（^C）
    MAP KEYS TERMINATED BY ':'
STORED AS TEXTFILE
LOCATION '/user/hadoop/employee';
```

查询雇员信息结果如图 2.2 所示。

```
hive> select * from employee_external;
OK
Michael ["Montreal","Toronto"]  {"sex":"Male","age":30} {"DB":80}       {"Product":["Developer","Lead"]}
Will    ["Montreal"]    {"sex":"Male","age":35} {"Perl":85}     {"Product":["Lead"],"Test":["Lead"]}
Shelley ["New York"]    {"sex":"Female","age":27}       {"Python":80}   {"Test":["Lead"],"COE":["Architect"]}
Lucy    ["Vancouver"]   {"sex":"Female","age":57}       {"Sales":89,"HR":94}    {"Sales":["Lead"]}
Time taken: 0.04 seconds, Fetched: 4 row(s)
```

图2.2　查询雇员信息

注意

默认情况下，分隔符的使用有两个关键点需要特别注意。

① ARRAY 中包含 ARRAY：外部的 ARRAY 以^B 分隔，内部嵌套的 ARRAY 以^C 分隔。

② MAP 值中包含 ARRAY：外部的 MAP 中的键值对之间以^B 分隔，一个键值对内的两个元素间以^C 分隔，部嵌套的 ARRAY 以^D 分隔。

（2）修改表

修改表包括重命名、添加列、更新列等操作。

表重命名的语法如下。

【语法】

```
ALTER TABLE table_name RENAME TO new_table_name。
```

示例 2-3

将雇员表“employee_external”改名为“employee_ext”。

关键代码：

```
hive> alter table employee_external rename to employee_ext;
```

添加和更新列的语法如下。

【语法】

```
ALTER TABLE table_name ADD|REPLACE
COLUMNS（col_name data_type[COMMENT col_comment],...）
```

示例 2-4

在雇员表“employee_ext”中增加一列“empid”。

关键代码：

```
hive> alter table employee_ext add columns(empid string);
```

注意

修改表时，ADD 代表新增一个字段，字段位置在所有列后面（partition 列前面）。REPLACE 表示替换表中所有字段。

3. DDL 其他常用命令

除了对数据库和表的创建和修改外，DDL 还包括其他常用操作，具体介绍如表 2-3 所示。

表 2-3　DDL 其他常用操作

命令	示例	说明
drop	drop database database_name;	删除数据库
	drop table table_name;	删除表
truncate	truncate table table_name;	清空表
show	show databases;	显示所有数据库名
	show tables;	列出当前数据库中所有表
	show views;	列出当前数据库中所有视图
	show partitions table_name;	列出表中所有分区
	show create table table_name;	显示建表语句
	show functions "a.*";	列出所有 UDF 和内置函数，支持通配符查找
describe（可简写为 desc）	describe database database_name;	显示数据库信息
	describe table_name;	显示表信息
	describe table_name.col_name;	显示表中某列的信息

详细 Hive DDL 操作演示视频请扫描二维码观看。

Hive DDL 操作

2.2.2　Hive DML 操作

从前面表操作部分的示例 2 中可以看到，创建外部表时没有进行数据插入（或者称数据装载），但是经过查询发现表中已有数据，这是因为在定义表的 Schema 时指定了“LOCATION”，此时只要该位置存在实际数据文件，Hive 将自动关联该目录下的数据文件。通常情况下，数据需要手动装载并进行维护管理，此时就需要使用 Hive DML 命令。

在 Hive 中，DML 操作包括 load、insert、update、delete、merge、import/export、explain 命令，分别对表进行数据装载、插入、更新、删除、合并、导入/导出以及执行计划查询等操作。

1. **数据装载与插入**

一般使用两种方式实现数据装载，分别是 load 和 insert 命令，二者有明显的区别。

➢ load 命令不对数据进行任何转换，只是简单地将数据复制或者移动至 Hive 表对应的位置。

➢ insert 命令将会执行 MapReduce 作业并将数据插入至 Hive 表中，一般使用较少。

（1）load 命令

【语法】

```
LOAD DATA [LOCAL] INPATH 'filepath'
[OVERWRITE] INTO TABLE table_name [PARTITION(partcol1=val1,partcol2=val2,...)]
```

➢ LOCAL：限制为本地文件系统路径，如果没有该关键字则是 HDFS 路径。

➢ OVERWRITE：覆盖目标文件夹中的数据，如果没有该关键字且目标文件夹中已存在同名文件，将保留之前的文件，新文件名后缀以自动序号区分。

➢ PARTITION：如果目标表是分区表，须使用该关键字为每个分区键指定一个值。

示例 2-5

根据示例 2-2 中 employee_external 的 Schema 创建内部表 emp，并使用 load 完成 emp 的数据装载。

分析：Hive 支持通过某个表的 Schema 来创建新表的 Schema。

关键代码：

```
hive> create table emp like employee_external;
hive> show create table emp;
```

新产生的表 emp 和 employee_external 有着相同的表结构与存储方式，但它们的 LOCATION 不同。下面进行手动数据装载。

```
hive> load data inpath '/user/hadoop/employee/employee.txt' into table emp;
    > select * from emp;
```

输出结果与示例 2-2 相同。

注意

如果是本地文件，则须将其拷贝至数据库表所在的位置；如果是 HDFS 文件则直接移动即可。

示例 2-6

创建雇员分区表 emp_partitioned，要求按照雇员的入职年份对表进行分区。

分析：

由于表中没有入职年份字段（假设 employee.txt 中的员工都是 2018 年入职），因此须首先创建基于入职年份（分区键为 year）的分区表，然后装载数据时指定分区键和值（year=2018）。

关键代码：

```
create table if not exists emp_partitioned (
    name string,
    work_place array<string>,
    sex_age struct<sex:string,age:int>,
    skills_score map<string,int>,
    depart_title map<string,array<string>>
)
--指定分区键“year”
partitioned by (year int)
row format delimited
    fields terminated by '|'
    collection items terminated by ','
    map keys terminated by ':'
stored as textfile;
--装载数据至表分区“year=2018”
load data
    inpath '/user/hadoop/employee/employee.txt'
    into table emp_partitioned partition(year=2018);
```

执行上述代码后，可查看 emp_partitioned 表的结构信息，如图 2.3 所示。

```
hive> desc emp_partitioned;
OK
name                    string
work_place              array<string>
sex_age                 struct<sex:string,age:int>
skills_score            map<string,int>
depart_title            map<string,array<string>>
year                    int

# Partition Information
# col_name              data_type               comment

year                    int
Time taken: 0.082 seconds, Fetched: 11 row(s)
```

图2.3　查看分区表结构信息

实际上，分区只是 HDFS 中表目录下对应的子目录。emp_partitioned 表对应的 HDFS 存储目录结构如图 2.4 所示。

Browse Directory

/home/hadoop/hive/warehouse/empdb.db/emp_partitioned/year=2018　Go!

Permission	Owner	Group	Size	Last Modified	Replication	Block Size	Name
-rwx-wx-wx	root	supergroup	227 B	Thu Dec 06 11:59:59 +0800 2018	3	128 MB	employee.txt

图2.4　分区表对应的HDFS存储目录结构

建立分区后，对表查询时在 where 子句中指定分区可提升查询效率。例如：

```
hive> select * from emp_partitioned where year=2018;
```

（2）insert 命令

【语法】

```
INSERT (INTO|OVERWRITE) TABLE table_name [PARTITION(partcol1=val1,partcol2=val2...)]
select_statement1 FROM from_statement;
```

示例 2-7

使用 insert 命令在 emp_partitioned 表中新增一个 2019 年入职的员工。

分析：

emp_partitioned 表已被定义为分区表，所以在使用 insert 命令插入员工信息时仅须指定分区键 year 的值即可。

关键代码：

```
hive> insert into table emp_partitioned partition(year=2019)
    > select 'Jason',array('Beijing'),named_struct('sex','Male','age',22),
    > map('Hive',90,'Hadoop',88),
    > map('Product',array('Developer'));
```

上述代码中，使用了 Hive 的内置函数 array()、named_struct()、map()分别创建 ARRAY、STRUCT 以及 MAP，目的是为了符合表中的列类型定义。另外，必须使用 partition 关键字指定分区，因为 emp_partitioned 表是分区表，其中的任何一条数据必须明确指定存储在哪个分区中。

执行上述代码所得结果如图 2.5 所示。

```
hive> select * from emp_partitioned;
OK
Michael ["Montreal","Toronto"]  {"sex":"Male","age":30} {"DB":80}        {"Product":["Developer","Lead"]}        2018
Will    ["Montreal"]    {"sex":"Male","age":35} {"Perl":85}     {"Product":["Lead"],"Test":["Lead"]}    2018
Shelley ["New York"]    {"sex":"Female","age":27}       {"Python":80}   {"Test":["Lead"],"COE":["Architect"]}   2018
Lucy    ["Vancouver"]   {"sex":"Female","age":57}       {"Sales":89,"HR":94}    {"Sales":["Lead"]}      2018
Jason   ["Beijing"]     {"sex":"Male","age":22} {"Hive":90,"Hadoop":88} {"Product":["Developer"]}       2019
Time taken: 0.054 seconds, Fetched: 5 row(s)
```

图2.5　insert命令执行结果

emp_partitioned 表中新的分区在 HDFS 中对应的目录结构如图 2.6 所示。

Permission	Owner	Group	Size	Last Modified	Replication	Block Size	Name
drwx-wx-wx	root	supergroup	0 B	Thu Dec 06 12:00:30 +0800 2018	0	0 B	year=2018
drwx-wx-wx	root	supergroup	0 B	Thu Dec 06 13:37:06 +0800 2018	0	0 B	year=2019

Browse Directory — /home/hadoop/hive/warehouse/empdb.db/emp_partitioned/ Go!

图2.6　新的分区在HDFS中对应的目录结构

新分区“year=2019”中的文件内容如图 2.7 所示。

```
[root@master hive]# hdfs dfs -cat /home/hadoop/hive/warehouse/empdb.db/emp_partitioned/year=2019/*
18/12/06 13:38:41 WARN util.NativeCodeLoader: Unable to load native-hadoop library for your platform
a classes where applicable
Jason|Beijing|Male,22|Hive:90,Hadoop:88|Product:Developer
```

图2.7　新分区中的文件内容

示例 2-8

创建雇员分桶表 emp_name_buckets，要求按员工姓名将所有员工分为两个桶。

分析：

创建表时指定分桶字段与数量即可，然后使用 insert 命令在表中插入数据。

关键代码：

```
create table if not exists emp_name_buckets (
    name string,
    work_place array<string>,
    sex_age struct<sex:string,age:int>,
    skills_score map<string,int>,
    depart_title map<string,array<string>>
)
--创建分桶
clustered by(name) into 2 buckets
row format delimited
    fields terminated by '|'
    collection items terminated by ','
    map keys terminated by ':'
stored as textfile;
--强制分桶
set hive.enforce.bucketing = true;
--使用 emp 表中的数据进行插入。
insert into table emp_name_buckets select * from emp;
```

执行上述代码可得分桶表数据，如图 2.8 所示。

```
hive> select * from emp_name_buckets;
OK
Shelley ["New York"]     {"sex":"Female","age":27}       {"Python":80}   {"Test":["Lead"],"COE":["Architect"]}
Will    ["Montreal"]     {"sex":"Male","age":35} {"Perl":85}     {"Product":["Lead"],"Test":["Lead"]}
Lucy    ["Vancouver"]    {"sex":"Female","age":57}       {"Sales":89,"HR":94}    {"Sales":["Lead"]}
Michael ["Montreal","Toronto"]  {"sex":"Male","age":30} {"DB":80}       {"Product":["Developer","Lead"]}
Time taken: 0.052 seconds, Fetched: 4 row(s)
```

图2.8　分桶表数据

表 emp_name_buckets 的分桶文件在 HDFS 中的目录结构如图 2.9 所示。

Browse Directory

/home/hadoop/hive/warehouse/empdb.db/emp_name_buckets　Go!

Permission	Owner	Group	Size	Last Modified	Replication	Block Size	Name
-rwx-wx-wx	root	supergroup	114 B	Thu Dec 06 14:42:20 +0800 2018	3	128 MB	000000_0
-rwx-wx-wx	root	supergroup	113 B	Thu Dec 06 14:42:21 +0800 2018	3	128 MB	000001_0

图2.9　分桶文件在HDFS中的目录结构

每个分桶文件中的内容如图 2.10 所示。

```
[root@master hive]# hdfs dfs -cat /home/hadoop/hive/warehouse/empdb.db/emp_name_buckets/000000_0
18/12/06 14:45:51 WARN util.NativeCodeLoader: Unable to load native-hadoop library for your platform...
a classes where applicable
Shelley|New York|Female,27|Python:80|Test:Lead,COE:Architect
Will|Montreal|Male,35|Perl:85|Product:Lead,Test:Lead
[root@master hive]# hdfs dfs -cat /home/hadoop/hive/warehouse/empdb.db/emp_name_buckets/000001_0
18/12/06 14:45:55 WARN util.NativeCodeLoader: Unable to load native-hadoop library for your platform...
a classes where applicable
Lucy|Vancouver|Female,57|Sales:89,HR:94|Sales:Lead
Michael|Montreal,Toronto|Male,30|DB:80|Product:DeveloperLead
```

图2.10　分桶文件内容

注意

分桶表插入数据前须先执行命令：set hive.enforce.bucketing = true;

2. 数据更新、删除与合并

从 0.14 版本起 Hive 支持数据更新与删除操作，从 2.2 版本起 Hive 支持数据合并操作。执行这些操作的表须在表上开启事务（ACID）支持，缺省情况下无须开启。Hive 是数据仓库解决方案，不适合进行更新、删除等事务操作。

3. 数据导入/导出

（1）export

【语法】

EXPORT TABLE table_name [PARTITION(partcol=partval,...)] TO 'export_target_path'

例如：

```
hive> export table emp to '/tmp/emp';
```

上述命令将导出 emp 表至 HDFS 目录/tmp/emp，包括其元数据与实际数据。在实际使用过程中，通常也利用 insert 命令完成数据的导出。

【语法】

INSERT OVERWRITE [LOCAL] DIRECTORY directory SELECT … FROM …

例如：

```
hive> insert overwrite local directory '/home/hadoop/hive/empdb.emp' select * from emp;
```

上述命令将导出 emp 表的数据至本地文件系统中。

（2）import

【语法】

IMPORT [[EXTERNAL] TABLE new_or_original_tablename [PARTITION(partcol1=val1,...)]]
FROM 'source_path' [LOCATION 'import_target_path']

例如：

```
hive> import table emp_new from '/tmp/emp';
```

上述代码将向 HDFS 目录/tmp/emp 导入数据并生成新的内部表 emp_new。

4. 数据执行计划查询

Hive 提供 explain 命令，可显示数据查询的执行计划。

【语法】

EXPLAIN [EXTENDED|AST|DEPENDENCY|AUTHORIZATION|LOCKS|VECTORIZATION] query

例如：

```
hive> explain select * from emp;
```

查询执行计划显示结果如图 2.11 所示。

```
hive> explain  select * from emp;
OK
STAGE DEPENDENCIES:
  Stage-0 is a root stage

STAGE PLANS:
  Stage: Stage-0
    Fetch Operator
      limit: -1
      Processor Tree:
        TableScan
          alias: emp
          Statistics: Num rows: 1 Data size: 227 Basic stats: COMPLETE Column stats: NONE
          Select Operator
            expressions: name (type: string), work_place (type: array<string>), sex_age (type: struct<sex:string,age:int>
), skills_score (type: map<string,int>), depart_title (type: map<string,array<string>>)
            outputColumnNames: _col0, _col1, _col2, _col3, _col4
            Statistics: Num rows: 1 Data size: 227 Basic stats: COMPLETE Column stats: NONE
            ListSink

Time taken: 0.061 seconds, Fetched: 17 row(s)
```

图2.11　查询执行计划

详细 Hive DML 操作演示视频请扫描二维码观看。

Hive DML 操作

2.2.3　Hive Shell

前面提到，无论是 CLI 还是 Beeline，Hive Shell 都可以通过两种模式运行，即交互式模式与命令行模式。开发测试时使用交互式模式较多，生产环境下通常使用命令行模式。本小节主要介绍 Hive 的命令行以及 Hive 参数配置方式。

1. Hive 命令行

【语法】

hive [-hiveconf x=y]* [<-i filename>]* [<-f filename>|<-e query-string>] [-S] [-d]

主要参数使用说明如下。

➢ -i：从文件初始化 HQL。

- -e：从命令行执行指定的 HQL。
- -f：执行 HQL 脚本，与-i 作用类似。
- -S：静默模式，在交互式模式中仅输出主要信息，即去掉“OK”和“Time taken...”。
- -d：自定义变量，以便在 HQL 脚本中使用。
- -hiveconf x=y：配置 Hive/Hadoop 参数。

示例 2-9

使用 Hive 命令行运行一个查询。

输入命令：

```
$ hive -e 'select count(1) from empdb.emp'
```

输出结果：

```
OK
4
Time taken:52.286 seconds,Fetched: 4 row(s)
```

示例 2-10

使用 Hive 命令行运行一个文件。

首先将 SQL 写在 query.hql 文件中。

```
use empdb;
select count(1) from emp;
```

然后使用“hive -f”执行脚本。

输入命令：

```
$ hive -f query.hql
```

输出结果：

```
OK
4
Time taken: 41.046 seconds,Fetched: 1 row(s)
```

示例 2-11

使用 Hive 命令行自定义变量，完成通用查询。

首先修改 query.hql 文件。

```
use ${db};
select count(1) from ${table};
```

然后使用“hive -f”执行脚本。

输入命令：

```
$ hive -f query.hql -d db=empdb -d table=emp
```

输出结果：

```
OK
4
Time taken: 30.318 seconds,Fetched: 1 row(s)
```

2. Hive 参数配置方式

开发 Hive 应用程序时，通常须手动修改 Hive 的参数。设定参数可以调优 HQL 代码的执行效率，或帮助解决定位问题。对于一般参数而言，有以下 3 种设定方式。

（1）配置文件设定参数

用户自定义配置文件：$HIVE_HOME/conf/hive-site.xml，全局永久配置，修改后需要重启 Hive 服务。

（2）命令行设定参数

启动 Hive 客户端（或其处于 Server 模式）时，可以通过在命令行添加-hiveconf param=value 来设定参数，例如：

```
$ hive -hiveconf hive.root.logger=INFO,console
```

这一设定只对本次启动的 Session 有效。如果以 Server 模式启动，则对所有请求的 Session 都有效，例如：

```
$ hive --service hiveserver2 --hiveconf hive.root.logger=INFO,console
```

（3）SET 关键字设定参数

可以在 HQL 中通过使用 SET 关键字来设定参数，这一设定的作用域也是 Session 级别的，例如：

```
set mapred.reduce.tasks=100;
```

上述 3 种设定方式的优先级依次递增，即 SET 关键字设定参数覆盖命令行设定参数，命令行设定参数又覆盖配置文件设定参数。

注意

某些系统级的参数，例如 log4j 相关的参数，必须采用前两种参数设定方式进行设定，因为那些参数的读取在 Session 建立以前已经完成了。

2.2.4 技能实训

使用 Hive Shell 命令行方式完成对 Hive 雇员招聘信息 employee_hr.txt 的管理操作，employee_hr.txt 的内容如图 2.12 所示。

```
Matias McGrirl|1|945-639-8596|2011-11-24
Gabriela Feldheim|2|706-232-4166|2017-12-16
Billy O'Driscoll|3|660-841-7326|2017-02-17
Kevina Rawet|4|955-643-0317|2012-01-05
Patty Entreis|5|571-792-2285|2013-06-11
Claudetta Sanderson|6|350-766-4559|2016-11-04
Bentley Oddie|7|446-519-0975|2016-05-02
Theressa Dowker|8|864-330-9976|2012-09-26
Jenica Belcham|9|347-248-4379|2011-05-02
Reube Preskett|10|918-740-2357|2015-03-26
Mary Skeldon|11|361-159-8710|2016-03-09
Ethelred Divisek|12|995-145-7392|2016-10-18
```

图2.12　employee_hr.txt

关键步骤如下。

（1）创建数据库“empdb”。

（2）在数据库“empdb”下新建分区表“emp_hr_partitioned”，其包含的字段有“name”、“employee_id”、“sin_number”（社保号）、“start_date”（入职时间），按照入职时间中的

年份（year）对分区表进行分区操作。

（3）使用 desc 查看表信息。

（4）通过查询语句向表中插入数据。

（5）使用 HQL 语句查询表中一共有多少条数据。

提示

使用下列语句可在表中插入一个分区。

```
insert into table emp_hr_partitioned partition(year=2018)
select * from emp_hr where year(start_date)=2018;
```

还可使用 Hive 动态分区完成表中分区的插入，详细操作请扫描二维码查看。

Hive 动态分区

任务 3　使用 Hive Java API 操作雇员表

【任务描述】

使用 Hive 提供的 Java API 实现对雇员表的操作。

【关键步骤】

（1）使用 Intellij IDEA 搭建项目工程。

（2）使用 Java API 完成对 Hive 雇员表的操作。

2.3.1　开发环境搭建

本节介绍使用 Java API 的方式来操作 Hive 中雇员表（empdb.emp），开发工具是 IntelliJ IDEA + Maven。

使用 IntelliJ IDEA+Maven 搭建开发环境的步骤如下。

（1）启动 HiveServer2。

```
$ hive --service hiveserver2 &
```

注意

这里启动的是 HiveServer2。早期由于 HiveServer 使用了 Thrift 接口，其不能处理多于一个客户端的并发请求，并且不能通过 HiveServer 的代码修正。因此在 0.11.0 版本中重写了 HiveServer 代码并得到了 HiveServer2，进而解决了上述问题。HiveServer2 支持多客户端的并发和认证，为开放 API 客户端（如 JDBC 和 ODBC）提供了更好的支持。

（2）在 IDEA 中创建 maven 项目 hive-javaapi。

（3）添加 Maven 依赖包 hive-jdbc。完整 Maven POM 文件如下。

```
<?xml version="1.0" encoding="UTF-8"?>
<project xmlns="http://maven.apache.org/POM/4.0.0"
         xmlns:xsi="http://www.w3.org/2001/XMLSchema-instance"
         xsi:schemaLocation="http://maven.apache.org/POM/4.0.0
                             http://maven.apache.org/xsd/maven-4.0.0.xsd">
    <modelVersion>4.0.0</modelVersion>
    <groupId>cn.kgc.bigdata.hive.chap02</groupId>
    <artifactId>hive-javaapi</artifactId>
    <version>1.0-SNAPSHOT</version>
    <repositories>
        <repository>
            <id>cloudera</id>
            <url>https://repository.cloudera.com/artifactory/cloudera-repos/</url>
        </repository>
    </repositories>
    <dependencies>
        <dependency>
            <groupId>org.apache.hive</groupId>
            <artifactId>hive-jdbc</artifactId>
            <version>1.1.0-cdh5.14.2</version>
        </dependency>
        <dependency>
            <groupId>junit</groupId>
            <artifactId>junit</artifactId>
            <version>4.11</version>
        </dependency>
    </dependencies>
</project>
```

2.3.2 JDBC 操作 Hive 数据库

JDBC 是一种用于执行 SQL 语句的 Java API，可以为不同数据库提供统一的、面向 Java 语言的访问接口。hive-jdbc 是专用于 Hive 的 JDBC 驱动，是提供 JDBC-Hive 通信的桥梁，前面已在 POM 中添加了该依赖。接下来对 Hive 数据库的操作均基于 JDBC 编程接口实现。

1. **创建数据库连接**

JDBC 连接数据库时通常需要提供 4 个必要参数：

- 驱动类；
- 连接地址；
- 用户名；
- 密码。

JDBC 连接 Hive 的驱动类为“org.apache.hive.jdbc.HiveDriver”，连接地址为“jdbc:hive2://HOSTNAME:10000”。默认情况下 HiveServer2 身份验证为 NONE（hive.server2.

authentication=NONE)，即不需要验证，所以此处暂不提供用户名和密码。详细的 HiveServer2 身份验证机制请扫描二维码获取。

HiveServer2 身份验证机制

示例 2-12

使用 JDBC 连接 Hive。

关键代码：

```
package cn.kgc.bigdata.hive.chap02;
import java.sql.*;
public class HiveHelper {
    private static String driverName="org.apache.hive.jdbc.HiveDriver";
    private static String url="jdbc:hive2://master:10000";
    private Connection conn=null;
    private Statement stmt=null;
    private ResultSet rs=null;
    /**
     * 获取连接
     * @return
     * @throws ClassNotFoundException
     * @throws SQLException
     */
    public Connection getConn() throws ClassNotFoundException, SQLException {
        if (null==conn)
        {
            Class.forName(driverName);
            conn=DriverManager.getConnection(url);
        }
        return conn;
    }
    /**
     * 关闭连接
     */
    public void close() {
        try {
            if (null != conn && !conn.isClosed())
                conn.close();
            } catch(SQLException e){
                e.printStackTrace();
        }finally {
            conn=null;
        }
    }
     //省略其他数据库操作方法
}
```

2．创建数据库

示例 2-13

在 Hive 中使用 JDBC 创建 empdb_jdbc 数据库。

关键代码：

```
/**
 * 创建数据库
 * @param dbName
 * @return
 */
public void createDatabase(String dbName){
        try {
                stmt=getConn().createStatement();
                stmt.execute("create database if not exists "+dbName);
        } catch (SQLException e) {
                e.printStackTrace();
        } catch (ClassNotFoundException e) {
                e.printStackTrace();
        }
}
```

3．创建表

示例 2-14

基于 employee.txt 文件中的数据结构，使用 JDBC 在 empdb_jdbc 数据库中创建 emp_new 表。

关键代码：

```
/**
 * 创建表
 * @param dbName
 * @param tableName
 * @return
 */
public void createTable(String dbName,String tableName){
        try {
                stmt=getConn().createStatement();
                stmt.execute("use "+dbName);
                String sql="create table if not exists "+tableName+" (\n" +
                                "       name string,\n" +
                                "       work_place array<string>,\n" +
                                "       sex_age struct<sex:string,age:int>,\n" +
                                "       skills_score map<string,int>,\n" +
                                "       depart_title map<string,array<string>>\n" +
                                ")\n" +
                                "row format delimited\n" +
                                "       fields terminated by '|'\n" +
```

```
                    "      collection items terminated by ','\n" +
                    "      map keys terminated by ':'\n" +
                    "stored as textfile";
            stmt.execute(sql);
        } catch (SQLException e) {
            e.printStackTrace();
        } catch (ClassNotFoundException e) {
            e.printStackTrace();
        }
}
/**
 * 获取指定数据库中所有表名
 * @param dbName
 * @return
 */
public void showTables(String dbName){
        try {
            stmt=getConn().createStatement();
            stmt.execute("use "+dbName);
            rs=stmt.executeQuery("show tables");
            while (rs.next()){
                System.out.println(rs.getString(1));
            }
        } catch (SQLException e) {
            e.printStackTrace();
        } catch (ClassNotFoundException e) {
            e.printStackTrace();
        }
}
```

4. 装载数据

示例 2-15

使用 JDBC 完成数据装载。

关键代码：

```
/**
 * 装载数据
 * @param localFile
 * @param tbName
 * @return
 */
public void loadData(String localFile,String tbName){
        String sql="load data local inpath '"+localFile+"' overwrite into table "+tbName;
        try {
            stmt=getConn().createStatement();
            stmt.execute(sql);
```

```
        } catch (SQLException e) {
            e.printStackTrace();
        } catch (ClassNotFoundException e) {
            e.printStackTrace();
        }
}
```

5. **查询数据**

示例 2-16

使用 JDBC 完成数据查询。

关键代码：

```
/**
 * 查询数据
 * @param tbName
 */
public void selectAll(String tbName){
        String sql="select * from "+tbName;
        try {
            stmt=getConn().createStatement();
            rs=stmt.executeQuery(sql);
            while (rs.next()){
                System.out.println(
                            rs.getString("name")+"\t"+
                            rs.getString("work_place")+"\t"+
                            rs.getString("sex_age")+"\t"+
                            rs.getString("skills_score")+"\t"+
                            rs.getString("depart_title")
                );
            }
        } catch (SQLException e) {
            e.printStackTrace();
        } catch (ClassNotFoundException e) {
            e.printStackTrace();
        }
}
//测试
public static void main(String[] args) {
        String dbName="empdb_jdbc";
        String tbName="emp_new";
        String tbFullName=dbName+"."+tbName;
        String localFile="/root/workspace/hive/employee.txt";
        HiveHelper helper=new HiveHelper();
        helper.createDatabase(dbName);
        helper.createTable(dbName,tbName);
        helper.showTables(dbName);
```

```
        helper.loadData(localFile, tbFullName);
        helper.selectAll(tbFullName);
    }
```

JDBC 操作 Hive 雇员表代码运行结果如图 2.13 所示。

```
Run: HiveHelper
"C:\Program Files\Java\jdk1.8.0_60\bin\java.exe" ...
18/12/07 11:40:28 INFO org.apache.hive.jdbc.Utils: Supplied authorities: master:10000
18/12/07 11:40:28 INFO org.apache.hive.jdbc.Utils: Resolved authority: master:10000
emp_new
Michael ["Montreal","Toronto"]  {"sex":"Male","age":30} {"DB":80}   {"Product":["Developer","Lead"]}
Will    ["Montreal"]    {"sex":"Male","age":35} {"Perl":85} {"Product":["Lead"],"Test":["Lead"]}
Shelley ["New York"]    {"sex":"Female","age":27}   {"Python":80}   {"Test":["Lead"],"COE":["Architect"]}
Lucy    ["Vancouver"]   {"sex":"Female","age":57}   {"Sales":89,"HR":94}    {"Sales":["Lead"]}

Process finished with exit code 0
```

图2.13　JDBC操作Hive雇员表

2.3.3　技能实训

使用 JDBC API 的方式完成雇员表创建、装载与查询操作。雇员表数据来自 employee_id.txt，该数据在 employee.txt 基础上增加了一列 employee_id，其表示雇员编号，要求按 employee_id 将雇员表分为两个桶。employee_id.txt 内容如图 2.14 所示。

```
Michael|100|Montreal,Toronto|Male,30|DB:80|Product:DeveloperLead
Will|101|Montreal|Male,35|Perl:85|Product:Lead,Test:Lead
Steven|102|New York|Female,27|Python:80|Test:Lead,COE:Architect
Lucy|103|Vancouver|Female,57|Sales:89,HR:94|Sales:Lead
Mike|104|Montreal|Male,35|Perl:85|Product:Lead,Test:Lead
Shelley|105|New York|Female,27|Python:80|Test:Lead,COE:Architect
Luly|106|Vancouver|Female,57|Sales:89,HR:94|Sales:Lead
Lily|107|Montreal|Male,35|Perl:85|Product:Lead,Test:Lead
Shell|108|New York|Female,27|Python:80|Test:Lead,COE:Architect
Mich|109|Vancouver|Female,57|Sales:89,HR:94|Sales:Lead
```

图2.14　employee_id.txt

关键步骤如下。

（1）创建 Maven 项目。

（2）通过 JDBC 方式连接 HiveServer2。

（3）使用 JDBC API 创建数据库。

（4）使用 JDBC API 创建表 emp_id_buckets。

（5）使用 JDBC API 对表进行数据装载、查询等操作。

本章小结

- Hive 的基本数据类型有整数、小数、文本、布尔及二进制、时间等类型。
- Hive 的复杂数据类型有 ARRAY、MAP、STRUCT 和 UNION。
- Hive 表的类型有内部表、外部表、分区表和分桶表。
- Hive 表的常用操作有表创建、数据装载与查询。

➢ 使用 Hive Java API 操作表前须首先启动 HiveServer2 服务。

本章作业

一、简答题

1．Hive 的复杂数据类型有哪些？

2．Hive 内部表和外部表的区别是什么？

二、编码题

1．使用 Hive Shell 的方式创建电影票信息表 ticket，字段包含：电影票编号、影片名称、时间、座位与价格。其中，座位与价格使用 STRUCT 类型。

2．将本地的数据文件装载到 ticket 中。

3．使用 HQL 语句查询 ticket 中的所有数据。

第 3 章

Hive 元数据

技能目标

- 理解 Hive 元数据的概念。
- 了解 Hive 元数据表的结构。
- 掌握 Hive 元数据的定义和操作。
- 理解 Hive 元数据的存储方式。

本章任务

任务 1　访问雇员数据的元数据信息。
任务 2　使用 Hive Java API 读取雇员表元数据。
任务 3　使用 HCatalog 管理雇员数据的元数据。

本章资源下载

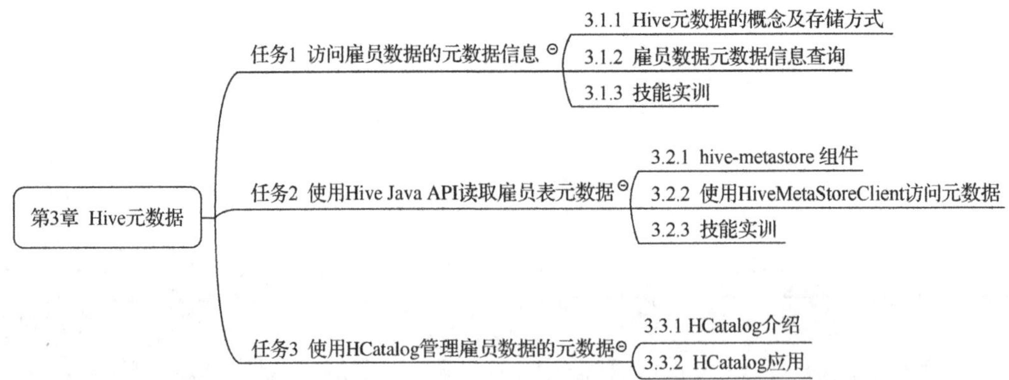

Hive 元数据是 Hive 中非常重要的概念。掌握并理解 Hive 元数据的概念和结构对于 Hive的应用十分重要。在实际的开发过程中，通过共享 Hive 元数据，其他应用也可以访问和操作 Hive 中的数据，这是常见的 Hive 元数据使用场景。通过本章的学习，读者须掌握定义和操作 Hive 元数据的方法。

任务 1 访问雇员数据的元数据信息

【任务描述】

了解 Hive 元数据的概念，熟悉常见的 Hive 元数据表结构以及存储方式。

【关键步骤】

（1）认识 Hive 元数据。

（2）了解 Hive 元数据表结构。

（3）查询雇员表的元数据信息。

3.1.1 Hive 元数据的概念及存储方式

前面章节提到 Hive 中有两类数据：真实数据和元数据。和关系型数据库一样，元数据可以看作是描述数据的数据，包括 Hive 表的数据库名、表名、字段名称与类型、分区字段与类型等。

Hive 将元数据存储在 RDBMS 中，有以下 3 种模式可以连接到数据库。

➢ Single User Mode：单用户模式，使用内置 Derby 数据库，也称内嵌 Derby 模式。

➢ Multi User Mode：多用户模式，使用本地 MySQL 数据库，也称本地模式。

➢ Remote Server Mode：远程服务模式，使用远程元数据服务访问数据库，也称远程模式。

按元数据存储位置划分，单用户与多用户模式均属于本地存储，远程服务模式属于远端存储。无论使用哪种模式，Hive 客户端均须首先连接 metastore 服务，然后由 metastore 服务去访问数据库以完成元数据的存取。

1. Single User Mode

默认情况下，Hive 使用 Derby 内存数据库保存元数据，优点是 Derby 小巧易用，安装非常方便。

使用 Derby 存储方式时，运行 Hive Shell 会在当前目录生成一个 derby.log 文件和一个 metastore_db 目录，用于保存一些 HQL 操作结果。这种存储方式的弊端是：

- 在同一个目录下同一时刻只能有一个 Hive 客户端使用数据库；
- 切换目录重新启动 Hive Shell 后，之前保存的元数据将无法被查看，即无法实现元数据共享。

鉴于上述弊端，我们将该模式称为单用户模式，一般在测试环境时使用（用于测试 Hive 环境）。在该模式下无须进行特别配置即可使用，但须注意默认的仓库地址（hive.metastore.warehouse.dir）为“/user/hive/warehouse”。

在单用户模式下 Hive 服务、metastore 服务、Derby 数据库运行在同一进程中，如图 3.1 所示。

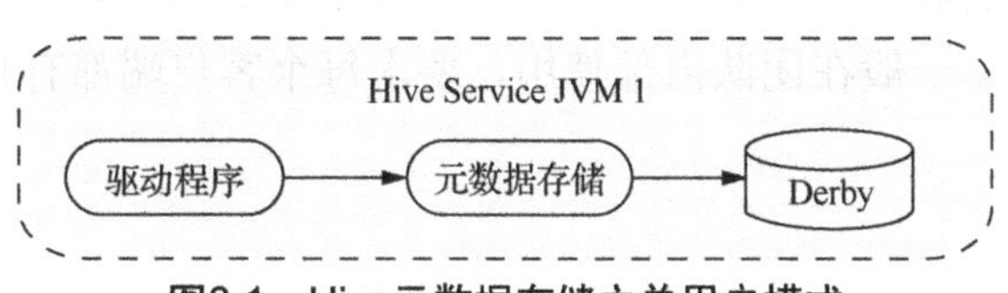

图3.1　Hive元数据存储之单用户模式

2. Multi User Mode

将元数据存储介质更换为 MySQL，即为多用户模式，这是开发中经常使用的模式。这种模式需要单独运行一个 MySQL 服务，并作如下配置（需要将 MySQL 的 JDBC 驱动 jar 文件拷贝到$HIVE_HOME/lib 目录下）。

（1）配置 MySQL

在 MySQL 中配置对应的 Hive 用户，例如，使用以下命令创建“hive”用户并授予权限。

```
create user 'hive'@'%' identified by 'hive';
grant all privileges on *.*  to 'hive'@'%' with grant option;
flush privileges;
```

通过以下命令可以查看 MySQL 的所有用户情况。

```
SELECT host,user,password FROM mysql.user;
```

（2）修改 hive-site.xml

关键代码：

```
<!—元数据存储配置-->
<property>
    <name>javax.jdo.option.ConnectionURL</name>
    <value>jdbc:mysql://MYSQL-HOST:3306/hive?createDatabaseIfNotExist=true</value>
</property>
<property>
```

```
        <name>javax.jdo.option.ConnectionDriverName</name>
        <value>com.mysql.jdbc.Driver</value>
    </property>
    <!-- Hive 连接 MySQL 的用户名和密码，用于访问元数据-->
    <property>
        <name>javax.jdo.option.ConnectionUserName</name>
        <value>hive</value>
    </property>
    <property>
        <name>javax.jdo.option.ConnectionPassword</name>
        <value>hive</value>
    </property>
```

该模式除了须使用 MySQL 作为元数据存储介质，它的其他特性和单用户模式相同。Hive 服务、metastore 服务运行在同一进程中，而 MySQL 服务运行在单独的进程中。该进程既可以在当前 Hive 主机中运行，也可以在其他远程主机中运行。多用户模式如图 3.2 所示。这种模式可以在不同主机上运行多个 Hive（作为 MySQL 客户端）并同时连接到 MySQL，一般在团队内部使用，要求每个客户端都有自己的 MySQL 用户名和密码。

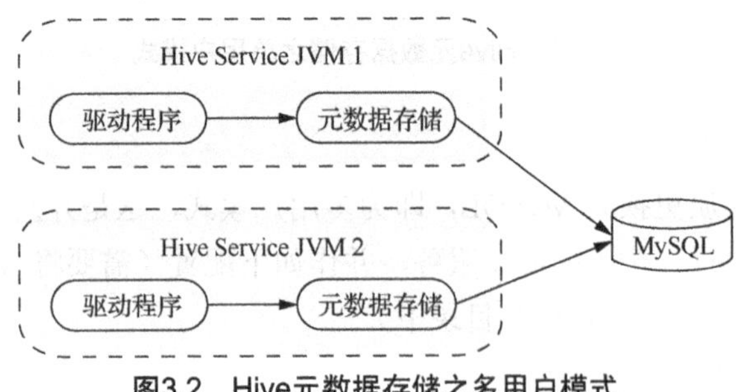

图3.2　Hive元数据存储之多用户模式

既使 MySQL 存储介质是其他主机，也不代表 Hive 元数据存储模式是远程模式。存储模式是否是远程模式须看 metastore 服务和 Hive 服务是否在同一进程中，即"远程"指的是远程的 metastore 服务而不是远程的 MySQL 服务。

3. Remote Server Mode

本地模式下，每个客户端可启动多个 Hive 副本，每个 Hive 都内置一个 metastore 服务，这样明显会浪费资源。所以可单独启动一个 metastore 服务，所有客户端使用 Thrift 协议通过该服务访问元数据库（如 MySQL），此模式称为远程模式。启动命令如下：

```
$ hive --service metastore -p 9083
```

其中"-p"用于指定监听端口，默认端口为 9083。

对 metastore 服务端来说，仍须指定 MySQL 连接信息以完成对 MySQL 的访问。

检查元数据服务是否启动成功的命令如下：

```
$ netstat -anp|grep 9083
```

9083 端口处于正常监听状态即可。

对 metastore 客户端来说，其无须再配置 MySQL 连接信息，指定 metastore 服务地址即可（主要是在 hive-site.xml 中配置 hive.metastore.uris）。这是一个类似 URL 的链接地址，用于通过 thrift 前缀连接 metastore。thrift 前缀的格式为：thrift://METASTORE-HOST:PORT，如果有多个 metastore 服务，须使用逗号将它们分隔。例如：

```
<property>
    <name>hive.metastore.uris</name>
    <value>thrift://192.168.9.80:9083,thrift://192.168.9.80:9084</value>
</property>
```

在远程模式下，Hive 服务和 metastore 服务可以在不同的进程中，也可以在不同的主机中，这样便可使二者解耦。所以在生产环境中，推荐使用远程模式访问元数据库。远程模式如图 3.3 所示。

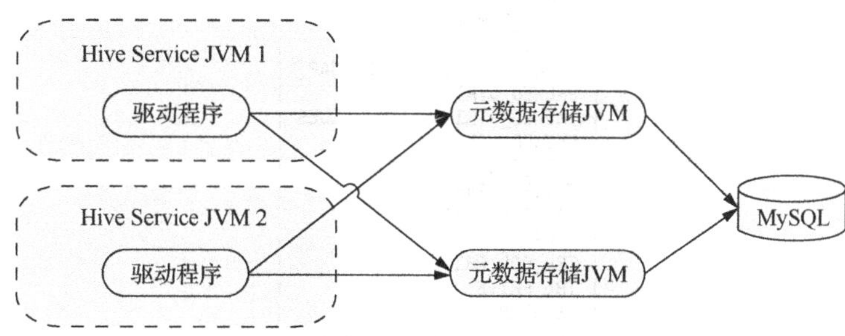

图3.3 Hive元数据存储之远程模式

3.1.2 雇员数据元数据信息查询

前面在 Hive 中创建过雇员表，接下来看一下雇员表的元数据在 MySQL 中是如何存储的。进入 MySQL，输入如下命令：

```
mysql> use hive;              --视 hive-site.xml 配置 javax.jdo.option.ConnectionURL 而定
mysql> show tables;
```

输出结果列出了 Hive 的所有元数据表名称，如图 3.4 所示。

其中，主要的元数据表介绍如下。

- VERSION：存储 Hive 版本。
- DBS、DATABASE_PARAMS：存储 Hive 数据库相关信息。
- TBLS、TABLE_PARAMS、TBL_PRIVS：存储 Hive 表相关信息。
- SDS、SD_PARAMS、SERDES、SERDE_PARAMS：存储 Hive 文件存储相关信息。
- PARTITIONS、PARTITION_KEYS、PARTITION_KEY_VALS、PARTITION_PARAMS：存储分区相关信息。
- COLUMNS_V2：存储表对应的字段信息。

```
+-----------------------------+
| Tables_in_hive              |
+-----------------------------+
| BUCKETING_COLS              |
| CDS                         |
| COLUMNS_V2                  |
| DATABASE_PARAMS             |
| DBS                         |
| DB_PRIVS                    |
| FUNCS                       |
| FUNC_RU                     |
| GLOBAL_PRIVS                |
| IDXS                        |
| INDEX_PARAMS                |
| NOTIFICATION_SEQUENCE       |
| PARTITIONS                  |
| PARTITION_KEYS              |
| PARTITION_KEY_VALS          |
| PARTITION_PARAMS            |
| PART_COL_PRIVS              |
| PART_COL_STATS              |
| PART_PRIVS                  |
| ROLES                       |
| SDS                         |
| SD_PARAMS                   |
| SEQUENCE_TABLE              |
| SERDES                      |
| SERDE_PARAMS                |
| SKEWED_COL_NAMES            |
| SKEWED_COL_VALUE_LOC_MAP    |
| SKEWED_STRING_LIST          |
| SKEWED_STRING_LIST_VALUES   |
| SKEWED_VALUES               |
| SORT_COLS                   |
| TABLE_PARAMS                |
| TAB_COL_STATS               |
| TBLS                        |
| TBL_COL_PRIVS               |
| TBL_PRIVS                   |
| VERSION                     |
+-----------------------------+
37 rows in set (0.00 sec)
```

图3.4　Hive所有元数据表名称

其他不常用的元数据表介绍如下。

- CDS：记录 COLUMN_V2 中的所有 CD_ID。
- DB_PRIVS：数据库权限信息表。
- IDXS：索引表。
- INDEX_PARAMS：索引相关属性信息。
- TBL_COL_STATS：表字段统计信息。
- TBL_COL_PRIVS：表字段授权信息。
- PART_PRIVS：分区授权信息。
- PART_COL_PRIVS：分区字段授权信息。
- PART_COL_STATS：分区字段的统计信息。
- FUNCS：用户注册函数。
- FUNC_RU：用户注册函数的资源信息。

下面来分析主要元数据表的详细结构。

1. VERSION

VERSION 表存储了 Hive 的版本信息，包括 ID 主键、Hive 版本及版本说明。VERSION

表中有且仅能有一条记录，否则 Hive 服务将无法启动。VERSION 表中的元数据如表 3-1 所示。

表 3-1　VERSION 元数据

VER_ID	SCHEMA_VERSION	VERSION_COMMENT
1	1.1.0-cdh5.14.2	Set by MetaStore root@192.168.9.80

2. ***数据库元数据***

（1）DBS

DBS 表存储 Hive 中所有数据库的基本信息。元数据表 DBS 字段说明如表 3-2 所示。

表 3-2　元数据表 DBS 字段说明

元数据表字段	说明	示例数据（以雇员数据库为例）
DB_ID	数据库 ID	26
DESC	数据库描述	雇员数据库
DB_LOCATION_URI	数据库 HDFS 路径	hdfs://cluster1/home/hadoop/hive/warehouse/empdb.db
NAME	数据库名	Empdb
OWNER_NAME	数据库所有者	Root
OWNER_TYPE	所有者类型	USER

（2）DATABASE_PARAMS

DATABASE_PARAMS 表存储数据库的相关参数，这些参数是在创建表时由“CREATE DATABASE…”的子句“WITH DBPROPERTIES(property_name=property_value,...)”指定的。例如：

```
create dabase testdb with dbproperties('createdby'='hive')
```

其中“testdb”在该元数据表中的示例如表 3-3 所示。

表 3-3　元数据表 DATABASE_PARAMS

元数据表字段	说明	示例数据（以雇员数据库为例）
DB_ID	数据库 ID	26
PARAM_KEY	参数名	createdby
PARAM_VALUE	参数值	root

其中，DBS 和元数据表 DATABASE_PARAMS 通过 DB_ID 字段关联。

3. ***表元数据***

TBLS、TABLE_PARAMS 和 TBL_PRIVS 表通过 TBL_ID 进行关联。

（1）TBLS

元数据表 TBLS 中存储 Hive 表的基本信息，包括内部表、外部表、索引表与视图。其详细结构如表 3-4 所示。

表 3-4　元数据表 TBLS

元数据表字段	说明	示例数据（以 empdb.emp 为例）
TBL_ID	表 ID	694
CREATE_TIME	创建时间	1544065674
DB_ID	数据库 ID	26
LAST_ACCESS_TIME	上次访问时间	0
OWNER	所有者	root
RETENTION	保留字段	0
SD_ID	序列化配置信息	694（对应 SDS 表中的 SD_ID）
TBL_NAME	表名	emp
TBL_TYPE	表类型	MANAGED_TABLE（内部表） 其他类型：EXTERNAL_TABLE（外部表）、INDEX_TABLE（索引表）、VIRTUAL_VIEW（视图）
VIEW_EXPANDED_TEXT	视图的详细 HQL 语句	NULL
VIEW_ORIGINAL_TEXT	视图的原始 HQL 语句	NULL

（2）TABLE_PARAMS

TABLE_PARAMS 表存储表/视图的额外属性信息，由建表语句的子句“TBLPROPERTIES (property_name= property_val)” 指定。其详细结构如表 3-5 所示。

表 3-5　元数据表 TABLE_PARAMS

元数据表字段	说明	示例数据（以 empdb.emp 为例）
TBL_ID	表 ID	694
PARAM_KEY	属性名	transient_lastDdlTime
PARAM_VALUE	属性值	1544065674

（3）TBL_PRIVS

TBL_PRIVS 表存储表/视图的授权信息，其详细结构如表 3-6 所示。

表 3-6　元数据表 TBL_PRIVS

元数据表字段	说明	示例数据（以 empdb.emp 为例）
TBL_GRANT_ID	授权 ID	1
CREATE_TIME	授权时间	1544065674
GRANT_OPTION	被授权者可授权给其他用户	0
GRANTOR	授权执行用户	root
GRANTOR_TYPE	授权者类型	USER
GRANTOR_NAME	被授权用户	username
PRINCIPAL_TYPE	被授权用户类型	USER
TBL_PRIV	权限	Select、Alter
TBL_ID	表 ID	694（对应 TBLS 表中的 TBL_ID）

4．表数据存储元数据

创建 Hive 表时可以指定各种文件的格式。Hive 将 HQL 解析成 MapReduce 时，须知道去哪里、使用哪种格式读写 HDFS 文件，而这些信息就保存在这几张表中。

（1）SDS

SDS 表保存文件存储的基本信息，如 INPUT_FORMAT、OUTPUT_FORMAT、是否压缩等。TBLS 表中的 SD_ID 与该表关联，可以通过 SD_ID 获取该表中保存的存储信息。元数据表 SDS 的详细结构如表 3-7 所示。

表 3-7　元数据表 SDS

元数据表字段	说明	示例数据
SD_ID	存储信息 ID	702
CD_ID	字段信息 ID	700（关联 COLUMN_V2 表中的 CD_ID）
INPUT_FORMAT	文件输入格式	org.apache.hadoop.mapred.TextInputFormat
IS_COMPRESSED	是否压缩	0
IS_STOREDASSUBDIRECTORIES	是否以子目录存储	0
LOCATION	HDFS 路径	Hdfs://cluster1/home/hadoop/hive/warehouse/empdb.db/emp_name_buckets
NUM_BUCKETS	分桶数量	2
OUTPUT_FORMAT	文件输出格式	org.apache.hadoop.hive.ql.io.HiveIgnoreKeyTextOutputFormat
SERDE_ID	序列化类 ID	702（对应 SERDES 表）

（2）SD_PARAMS

SD_PARAMS 表存储 Hive 存储的额外属性信息，其详细结构如表 3-8 所示。

表 3-8　元数据表 SD_PARAMS

元数据表字段	说明	示例数据
SD_ID	存储配置 ID	1
PARAM_KEY	存储属性名	test_name
PARAM_VALUE	存储属性值	test_value

（3）SERDES

SERDES 表存储序列化使用的类信息，在创建表时由子句“ROW FORMAT SERDE 'org.apache.hadoop.hive.serde2.lazy.LazySimpleSerde '”指定。其详细结构如表 3-9 所示。

表 3-9　元数据表 SERDES

元数据表字段	说明	示例数据
SERDE_ID	序列化类配置 ID	694
NAME	属性名	NULL
SLIB	序列化类	org.apache.hadoop.hive.serde2.lazy.LazySimpleSerde

（4）SERDE_PARAMS

SERDE_PARAMS 表存储序列化的属性和格式信息，如行、列分隔符等，这些信息会在创建表时由“ROW FORMAT SERDE '……' WITH SERDEPROPERTIES(property_name=property_val)”指定。元数据表 SERDE_PARAMS 的详细结构如表 3-10 所示。

表 3-10　元数据表 SERDE_PARAMS

元数据表字段	说明	示例数据
SERDE_ID	序列化类配置 ID	694
PARAM_KEY	属性名	field.delim
PARAM_VALUE	属性值	|

5. *表分区元数据*

存储表分区相关信息的元数据表主要有 PARTITIONS、PARTITION_KEYS、PARTITION_KEY_VALS 和 PARTITION_PARAMS。

（1）PARTITIONS

PARTITIONS 表存储表分区的基本信息，其详细结构如表 3-11 所示。

表 3-11　元数据表 PARTITIONS

元数据表字段	说明	示例数据（以 emp_partitioned 表为例）
PART_ID	分区 ID	1
CREATE_TIME	分区创建时间	1544068830
LAST_ACCESS_TIME	最后一次访问时间	0
PART_NAME	分区名	year=2018
SD_ID	分区存储信息 ID	696
TBL_ID	表 ID	695

（2）PARTITION_KEYS

PARTITION_KEYS 表存储分区的字段信息，其详细结构如表 3-12 所示。

表 3-12　元数据表 PARTITION_KEYS

元数据表字段	说明	示例数据（以 emp_partitioned 表为例）
TBL_ID	表 ID	695
PKEY_COMMENT	分区字段说明	NULL
PKEY_NAME	分区字段名	year
PKEY_TYPE	分区字段类型	int
INTEGER_IDX	分区字段顺序	0

（3）PARTITION_KEY_VALS

PARTITION_KEY_VALS 表存储分区字段值，其详细结构如表 3-13 所示。

表 3-13　元数据表 PARTITION_KEY_VALS

元数据表字段	说明	示例数据（以 emp_partitioned 表为例）
PART_ID	分区 ID	1
PART_KEY_VAL	分区字段值	2018
INTEGER_IDX	分区字段值顺序	0

（4）PARTITION_PARAMS

PARTITION_PARAMS 表存储分区的属性信息，其详细结构如表 3-14 所示。

表 3-14　元数据表 PARTITION_PARAMS

元数据表字段	说明	示例数据
PART_ID	分区 ID	2
PARAM_KEY	分区属性名	numFiles、numRows
PARAM_VALUE	分区属性值	1、0

6. 表字段元数据

COLUMNS_V2 表存储表对应的字段信息，其详细结构如表 3-15 所示。

表 3-15　元数据表 COLUMNS_V2

元数据表字段	说明	示例数据（以 emp 表为例）
CD_ID	字段信息 ID	694（所有 CD_ID 均会被 CDS 表记录）
COMMENT	字段注释	NULL
COLUMN_NAME	字段名	depart_title
TYPE_NAME	字段类型	map<string,array<string>>
INTEGER_IDX	字段顺序	4

示例 3-1

通过元数据查询 Hive 雇员表中的所有字段信息。

分析：

已知 Hive 雇员表的数据库为“empdb”，表名为“emp”，则可首先由 TBL 得到 SD_ID，然后在 SDS 表中由 SD_ID 得到 CD_ID，最后在 COLUMNS_V2 表中由 CD_ID 得到所有字段信息。

实现步骤：

（1）从 DBS 获得 DB_ID；

（2）从 TBLS 获得 SD_ID；

（3）从 SDS 获得 CD_ID；

（4）从 COLUMNS_V2 获得字段列表。

关键代码：

```
mysql> select * from COLUMNS_V2 where CD_ID in
              (select CD_ID from SDS where SD_ID in
```

```
    (select SD_ID from TBLS where DB_ID in
        (select DB_ID from DBS where NAME='empdb')
        and TBL_NAME='emp'));
```

使用元数据查询表结构的代码输出结果如图 3.5 所示。

```
+-------+---------+--------------+-----------------------------+--------------+
| CD_ID | COMMENT | COLUMN_NAME  | TYPE_NAME                   | INTEGER_IDX |
+-------+---------+--------------+-----------------------------+--------------+
|   694 | NULL    | depart_title | map<string,array<string>>   |            4 |
|   694 | NULL    | name         | string                      |            0 |
|   694 | NULL    | sex_age      | struct<sex:string,age:int>  |            2 |
|   694 | NULL    | skills_score | map<string,int>             |            3 |
|   694 | NULL    | work_place   | array<string>               |            1 |
+-------+---------+--------------+-----------------------------+--------------+
```

图3.5　使用元数据查询表结构

3.1.3　技能实训

搭建 CDH Hive 环境，新建数据库“hive_db”。在该数据库下新建表“student”，并使其包含字段：Sno、Sname、Sex、Sage、Sdept。在元数据表 DBS 中查询数据库“hive_db”的相关信息。并在元数据表 TBLS 中查询表“student”的元数据信息。

关键步骤如下。

（1）启动 Hadoop 集群，启动 MySQL。

（2）新建数据库“hive_db”，新建表“student”。

（3）使用 SQL 语句查询相关元数据的信息。

任务 2　使用 Hive Java API 读取雇员表元数据

【任务描述】

使用 HiveMetaStoreClient 管理 Hive 元数据。

【关键步骤】

（1）了解 hive-metastore 组件。

（2）使用 HiveMetaStoreClient 访问 Hive 雇员表元数据。

3.2.1　hive-metastore 组件

Hive 为 Java 开发环境提供了如下 2 种方式以对元数据进行访问。

（1）hive-jdbc：通过 JDBC 方式连接 HiveServer2 实现，在前面章节中已讲解过。

（2）hive-metastore：通过访问 metastore 服务实现，必须启动 metastore 服务以使元数据通过远程模式存储。

如果仅获取 Hive 元数据信息，而不操作实际数据，使用 hive-metastore 组件更加合适。其 JAR 文件位于：$HIVE_HOME/lib/hive-metastore-1.1.0-cdh5.14.2.jar，原理是通过

使用面向对象的方式访问数据库数据（Java Data Objects，JDO），进而实现对数据库中的元数据进行操作。hive-metastore 的组件结构如图 3.6 所示。

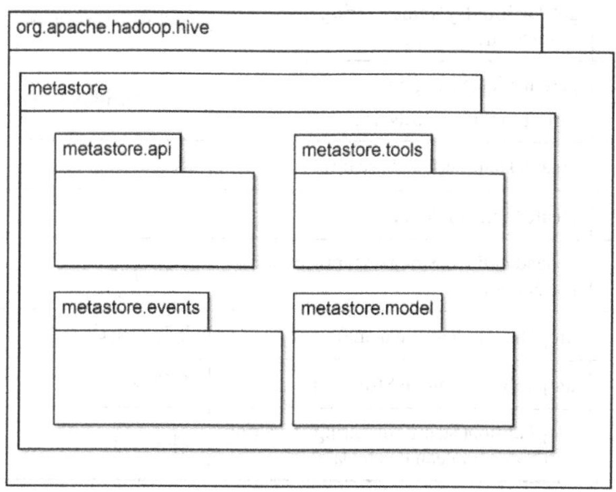

图3.6 hive-metastore组件结构

从图 3.6 中可以看出，hive-metastore 主要包括 5 个包。

（1）metastore：包含模块入口，是整个 hive-metastore 组件的核心。metastore 包含 HiveMetaStoreClient 等类。

（2）metastore.api：包含调用 hive-metastore 的接口。客户端须使用该包下的 API 进行元数据访问，其包含 Database、Table、Partition 及 FieldSchema 等类。

（3）metastore.events：包含事件处理与通知机制。当对元数据进行操作时会产生“event”（事件），它们会被“listener”（观察者）捕获并进行相应处理。

（4）metastore.model：包含用于数据持久化和对元数据表进行映射的类，例如将 DBS 表映射为 MDatabase 类，且可在 hive-metastore JAR 文件的根目录中找到 package.jdo 文件查看映射关系。

（5）metastore.tools：包含元数据管理员管理元数据时使用的工具。

为了完成对雇员表元数据的获取，使用 HiveMetaStoreClient 即可。HiveMetaStoreClient 实现了 IMetaStoreClient 接口，其包含的主要方法如表 3-16 所示。

表 3-16 IMetaStoreClient 接口主要方法

返回类型	方法	描述
List<String>	getAllDatabases()	从元数据中获取所有数据库名
Database	getDatabase(String dbName)	获得 Database 对象
List<String>	getAllTables(String dbName)	从指定的数据库获取所有表名
Table	getTable(String db,String tb)	获得 Table 对象
List<FieldSchema>	getSchema(String db,String tb)	获取指定表的所有字段信息
List<Partition>	listPartitions(String db,String tb,short max_parts)	获取分区列表

续表

返回类型	方法	描述
List<Partition>	getPartitionByNames(String db,String tb,List<String> partNames)	按分区名称获取所有分区
Partition	getPartition(String db, String tb,String partName)	获得 Partition 对象
void	createDatabase(Database db)	创建数据库
void	createTable(Table tbl)	创建表
Partition	appendPartition(String db,String tb,String partName)	添加分区
void	dropDatabase(String name)	删除数据库
void	dropTable(String db,String tb)	删除表
boolean	dropPatition(String db,String tb,String partName,boolean deleteData)	删除分区

Metastore API 详解请扫描二维码查看。

Metastore API 详解

3.2.2 使用 HiveMetaStoreClient 访问元数据

下面通过示例来了解 HiveMetaStoreClient 的使用方法。首先在 Intellij IDEA 中创建一个新的 Maven 项目，并添加如下依赖：

```
<dependency>
    <groupId>org.apache.hive</groupId>
    <artifactId>hive-metastore</artifactId>
    <version>1.1.0-cdh5.14.2</version>
</dependency>
<dependency>
    <groupId>org.apache.hadoop</groupId>
    <artifactId>hadoop-client</artifactId>
    <version>2.6.0-cdh5.14.2</version>
</dependency>
```

其中 hive-metastore 模块依赖 Hadoop 的 Configuration、JobContext 等类，所以须在其中加入 hadoop-client 模块。

确认 metastore 服务已启动，使用后台进程方式运行。

```
$ hive --service metastore &
```

1. 连接 metastore 服务

示例 3-2

获得 HiveMetaStoreClient 的实例。

分析：

须使用 HiveConf 来配置 hive.metastore.uris 的属性。HiveConf 继承自 Hadoop 的 Configuration，所以其与 Hadoop、HBase 类似。此处同样支持下列两种配置方式。

➢ 使用 HiveConf 的 set(String name,String value)方法。

➢ 使用 HiveConf 的 addResource(String file)加载类路径下的 hive-site.xml。

此外我们选择后者。

关键代码：

```
<!--hive-site.xml-->
<configuration>
    <!—配置 metastore 服务地址-->
    <property>
        <name>hive.metastore.uris</name>
        <value>thrift://192.168.9.80:9083</value>
    </property>
</configuration>
//HiveClient.java
import org.apache.hadoop.hive.conf.HiveConf;
import org.apache.hadoop.hive.metastore.HiveMetaStoreClient;
import org.apache.hadoop.hive.metastore.api.FieldSchema;
import org.apache.hadoop.hive.metastore.api.MetaException;
import org.apache.hadoop.hive.metastore.api.Partition;
import org.apache.thrift.TException;
import java.util.Arrays;
import java.util.List;
public class HiveClient {
    HiveMetaStoreClient client;
    /**
     * 初始化客户端
     */
    public HiveClient() {
        try {
            HiveConf conf=new HiveConf();
            conf.addResource("hive-site.xml");
            client=new HiveMetaStoreClient(conf);
        } catch (MetaException e) {
            e.printStackTrace();
        }
    }
    //省略其他元数据访问功能代码
}
```

当 HiveClient 的实例对象被创建时，须完成到 metastore 服务的连接。

注意

代码中的 hive-site.xml 放在 src/resources 目录下，resources 为资源根目录。

2. 获取数据库名称

示例 3-3

获取元数据中的所有数据库名称列表。

关键代码：

```
/**
* 返回所有数据库名称
* @return
*/
public List<String> getAllDatabases(){
	List<String> databases = null;
	try {
		databases = client.getAllDatabases();
	} catch (MetaException e) {
		e.printStackTrace();
	}
	return databases;
}
```

代码中的“client”为示例 3-2 中的 HiveMetaStoreClient 实例。

3. 获取表名

示例 3-4

获取指定数据库中所有表名。

关键代码：

```
/**
* 返回数据库中所有表名
* @param db
* @return
*/
public List<String> getAllTables(String db) {
	List<String> tables = null;
	try {
		tables = client.getAllTables(db);
	} catch (TException e) {
		e.printStackTrace();
	}
	return tables;
}
```

4. 获取字段

示例 3-5

获取指定表的所有字段信息。

关键代码：

```
/**
* 获取表中所有字段
```

```
* @param db
* @param table
* @return
*/
public List<FieldSchema> getSchema(String db, String table) throws Exception {
    List<FieldSchema> schema = null;
    try {
      if(client.tableExists(db,table))
        schema = client.getSchema(db, table);
    else
        throw new Exception("Table not found");
    } catch (TException e) {
        e.printStackTrace();
    }
    return schema;
}
```

5. 获取 HDFS 位置

示例 3-6

获取指定表对应的 HDFS 存储位置。

分析：

由于无法直接使用 HiveMetaStoreClient 来获取 HDFS 存储位置，故须通过 Table 实例的 StoreageDescriptor 属性来间接获取，这类似于在 MySQL 中由 TBLS 表得到 SD_ID，然后再通过查询 SDS 表得到 LOCATION。

关键代码：

```
/**
* 获取表对应的 HDFS 存储位置
* @param db
* @param table
* @return
*/
public String getLocation(String db, String table) throws Exception {
     String location = null;
     try {
         if(client.tableExists(db,table))
               location = client.getTable(db, table).getSd().getLocation();
         else
               throw new Exception("Table not found");
     }catch (TException e) {
         e.printStackTrace();
     }
     return location;
}
```

6. 获取分区

示例 3-7

获取分区表的分区信息。

关键代码：

```
/**
* 获取分区表的分区信息，已知分区名称时使用
* @param db
* @param table
* @param partNames 分区名称列表
* @return 返回 Partition 实例列表
*/
public List<Partition> getPartitions(String db,String table,List<String> partNames){
    List<Partition> partitions=null;
    try {
        partitions=client.getPartitionsByNames(db,table,partNames);
    } catch (TException e) {
        e.printStackTrace();
    }
    return partitions;
}
/**
* 获取分区表的所有分区
* @param db
* @param table
* @param maxParts
* @return 返回 Partition 实例列表
*/
public List<Partition> listPartitions(String db,String table,short maxParts){
    List<Partition> partitions=null;
    try {
        partitions=client.listPartitions(db,table,maxParts);
    } catch (TException e) {
        e.printStackTrace();
    }
    return partitions;
}
/**
* 获取分区表的所有分区名称
* @param db
* @param table
* @param maxParts
* @return 返回分区名称
*/
public List<String> listPartitionNames(String db,String table,short maxParts){
```

```
        List<String> partitionNames=null;
        try {
            partitionNames=client.listPartitionNames(db,table,maxParts);
        } catch (TException e) {
            e.printStackTrace();
        }
        return partitionNames;
    }
```

7．代码测试

示例 3-8

使用 main()方法完成对示例 2-8 中的代码测试。

关键代码：

```
/**
 * 测试
 * @param args
 */
public static void main(String[] args) {
    HiveClient cli=new HiveClient();
    System.out.println("======数据库列表======");
    for(String dbname : cli.getAllDatabases()){
        System.out.println(dbname);
    }
    String db="empdb",table="emp";
    System.out.println("======数据库"+db+"中所有表======");
    for(String tbname:cli.getAllTables(db)){
        System.out.println(tbname);
    }
    System.out.println("======数据库"+db+"中"+table+"表的所有字段======");
    try {
        for(FieldSchema field:cli.getSchema(db,table)){
            System.out.println(
                        String.format("%-15s\t%-30s\t%-50s",
                                field.getName(),
                                field.getType(),
                                field.getComment()));
        }
    } catch (Exception e) {
        e.printStackTrace();
    }
    System.out.println("======数据库"+db+"中"+table+"表 HDFS 位置======");
    try {
        System.out.println(cli.getLocation(db,table));
    } catch (Exception e) {
        e.printStackTrace();
```

```
        }
        table="emp_partitioned";
        System.out.println("======数据库"+db+"中"+table+"表分区======");
        for(String partName:cli.listPartitionNames(db,table,(short)5)){
            System.out.println(partName);
        }
}
```

使用 HiveMetaStoreClient 访问元数据程序的运行结果如图 3.7 所示。

```
======数据库列表======
default
empdb
empdb_jdbc
job
kdc
nginx
result
test
test2018121
======数据库empdb中所有表======
emp
emp_name_buckets
emp_partitioned
employee_external
======数据库empdb中emp表的所有字段======
name            string                              null
work_place      array<string>                       null
sex_age         struct<sex:string,age:int>          null
skills_score    map<string,int>                     null
depart_title    map<string,array<string>>           null
======数据库empdb中emp表HDFS位置======
hdfs://cluster1/home/hadoop/hive/warehouse/empdb.db/emp
======数据库empdb中emp_partitioned表分区======
year=2018
year=2019
```

图3.7　使用HiveMetaStoreClient访问元数据

更多 HiveMetaStoreClient 的使用请扫描二维码查看。

HiveMetaStoreClient 的使用

3.2.3　技能实训

编写 Java 代码，使用 HiveMetaStoreClient 完成对“hive_db.student”表的元数据查询。

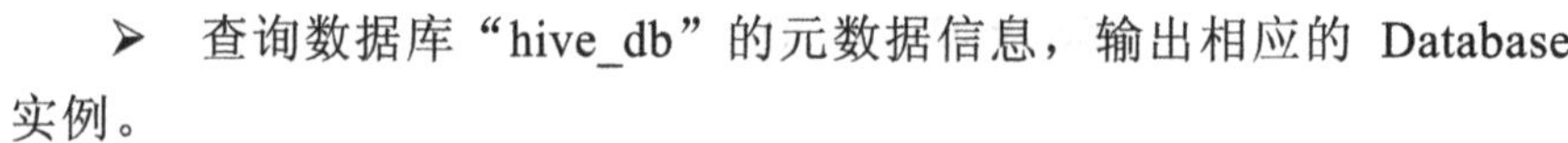

➢　查询数据库“hive_db”的元数据信息，输出相应的 Database 实例。

➢　查询表“student”的元数据信息，输出相应的 Table 和 FieldSchema 实例。

关键步骤如下。

（1）启动 metastore 服务。

（2）创建 HiveMetaStoreClient 实例并连接 metastore 服务。

（3）使用 getDatabase()方法获取 Database 实例。

（4）使用 getTable()方法获取 Table 实例。

（5）使用 getSchema()方法获取 FieldSchema 实例。

任务 3 使用 HCatalog 管理雇员数据的元数据

【任务描述】

使用 HCatalog 管理 Hive 元数据。

【关键步骤】

（1）认识 HCatalog。

（2）使用 HCatalog 操作元数据。

3.3.1 HCatalog 介绍

1. 基本概念

使用 Hive 处理数据时，相比于 MapReduce 编程，用户无须关注实际数据的存储位置以及存储格式，这样的好处是数据的生产与消费相互分离。除了 Hive，Hadoop 生态系统中还有其他与 Hive 类似的数据处理工具，如 Impala、Pig 等，每一个工具都有自己的优势，用户往往会针对不同的任务选择最合适的工具。

共享同一个元数据库可以使用户在多种工具之间更容易地共享数据。实际上，Impala 就可以完全使用 Hive 的元数据库。Pig 虽然不依赖 Hive 元数据，但是如果其能够首先对数据进行统一的抽取、转换、装载（Extract-Transform-Load，ETL），然后使用 Hive 进行数据分析，这样既可利用 Pig 的 ETL 优势，又可利用 Hive 的分析查询优势。在 Impala、Pig、Hive 等工具共用同一个元数据库时，每种工具都可以访问到其他工具生成的数据，这样便可减少不同工具单独装载和转换的步骤。

HCatalog 便是这样一个共享元数据的工具，是 Apache 开源的表存储管理工具，它将 Hive 元数据暴露给其他应用程序，允许不同的数据处理工具像二维表一样操作 Hive 元数据，而不用担心数据的格式与存储位置。HCatalog 底层依赖 Hive 元数据，其在执行操作过程中会创建一个 HiveMetaStoreClient 的实例，通过该实例的 API 来获取表的元数据信息。

从 0.11 版本之后 Hive 已经集成了 HCatalog，其所在目录位置为：$HIVE_HOME/hcatalog。

2. 架构

HCatalog 构建在 Hive 元数据上，并整合了 Hive 的 DDL。HCatalog 为 Pig 和 MapReduce 提供了读写接口，并且提供了命令行界面（Command Line Interface，CLI）以使用 Hive 的命令行接口来发布数据定义和元数据探索命令。注意 HCatalog 的 CLI 支持所有 Hive 而不会产生 MapReduce 的 DDL。HCatalog 主要提供了以下 3 类接口。

（1）HCatLoader 与 HCataStore：用于 Pig，使 Pig 只须知道表名就可以访问数据。

（2）HCatInputFormat 和 HCatOutputFormat：为 MapReduce 程序提供输入/输出接口。

（3）HCatReader 和 HCatWriter：提供不使用 MapReduce 即可进行并行输入/输出的数据传输 API。

HCatalog 的架构如图 3.8 所示。

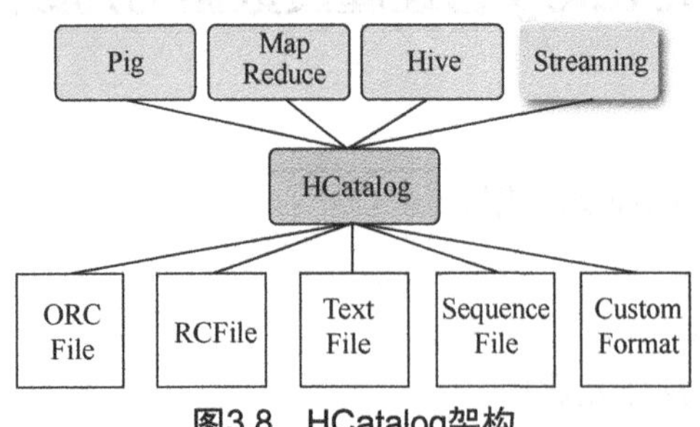

图3.8　HCatalog架构

3.3.2　HCatalog 应用

1．HCatalog CLI

因为 HCatalog 使用的是 Hive 的元数据存储，所以 Hive 用户不需要使用其他额外的工具来访问元数据，它们可以像以前一样使用 Hive 命令行工具。而对于非 Hive 用户而言，HCatalog 提供了 hcat 命令行工具，这个工具和 Hive 的命令行工具有些类似，两者最大的不同是 hcat 只接受不会产生 MapReduce 作业的命令，这意味着其可以支持大部分的 DDL 语句，如创建表等操作。hcat 命令行用法如表 3-17 所示。

表 3-17　hcat 命令行用法

选项	描述	例子
-e	通过命令行执行 DDL 语句	hcat -e "show tables;"
-f	执行包含 DDL 语句的脚本文件	hcat -f setup.hql
-g	为创建的表指定组	hcat -g mygroup ...
-p	为创建的表指定目录权限	hcat -p rwxr-xr-x
-D	将键值对以 Java 系统属性的形式传递给 HCatalog	hcat -Dlog.level=INFO
-h	提示用法帮助信息	hcat -h 或者 hcat

注意

使用 hcat 命令行工具前须将$HIVE_HOME/hcatalog/bin 目录追加到$PATH 变量中。

hcat 命令行不支持如下这些 HQL 操作：

- SELECT
- CREATE TABLE AS SELECT
- INSERT
- LOAD

- ALTER INDEX REBUILD
- ALTER TABLE CONCATENATE
- ALTER TABLE ARCHIVE
- ANALYZE TABLE
- EXPORT TABLE
- IMPORT TABLE

示例 3-9

使用 hcat 查询雇元数据的元数据。

关键代码：

```
$ cd $HIVE_HOME/hcatalog/bin
$ hcat -e "use empdb;show tables;"
```

输出结果：

```
OK
emp
emp_name_buckets
emp_partitioned
employee_external
Time taken: 0.256 seconds
```

示例 3-10

使用 hcat 查看 emp 表的结构脚本。

关键代码：

```
$ vi descemp.hql
--HQL 语句
use empdb;
desc emp;
$ hcat -f "descemp.hql"
```

输出结果：

```
OK
name                    string
work_place                      array<string>
sex_age                         struct<sex:string,age:int>
skills_score                    map<string,int>
depart_title                    map<string,array<string>>
Time taken: 0.362 seconds
```

示例 3-11

使用 hcat 基于 emp 表创建新的雇员表 emp_id，并增加字段 id。

```
$ vi create_emp_id.hql
--HQL 语句
use empdb;
create table if not exists emp_id like emp;
alter table emp_id add columns(id int);
```

```
desc emp_id;
$ hcat -f "create_emp_id.hql"
```

输出结果：

```
OK
name                    string
work_place              array<string>
sex_age                 struct<sex:string,age:int>
skills_score            map<string,int>
depart_title            map<string,array<string>>
id                      int
Time taken: 0.186 seconds
```

示例 3-12

使用 hcat 查看 emp_partitioned 表的分区。

```
$ hcat -e "use empdb;show partitions emp_partitioned;"
```

输出结果：

```
OK
year=2018
year=2019
Time taken: 0.336 seconds
```

2. HCatalog MapReduce API

HCatalog 为多个工具提供了 API，此处仅介绍 HCatInputFormat 与 HCatOutputFormat 的使用。二者使用 MapReduce 作业分别进行 HCatalog 管理表中数据的读取以及将处理以后的数据再写入 HCatalog 管理表中。

HCatInputFormat 使用 MapReduce 作业读取 HCatalog 管理表中的数据所使用的主要方法介绍如下。

- public static void setInput(Job job,String databaseName,String tableName)：为作业设置输入信息，将分区序列设置为 MapReduce 作业配置。
- public static void setOutputSchema(Job job,HCatSchema schema)：为作业设置由 HCatInputFormat 返回的 HCatRecorder 结构，其中 HCatRecorder 是 HCatalog 提供的一个用于记录和交互的类。
- public static HCatSchema getTableSchema(JobContext context)：获取 HCatTable 结构。

示例 3-13

使用 HCatInputFormat 读取 empdb.emp 表中雇员的年龄信息。

关键代码：

```
pom.xml
<project>
    <!—省略其他 POM 配置-->
    <dependencies>
        <dependency>
```

```
            <groupId>org.apache.hive.hcatalog</groupId>
            <artifactId>hive-hcatalog-core</artifactId>
            <version>1.1.0-cdh5.14.2</version>
        </dependency>
        <dependency>
            <groupId>org.apache.hadoop</groupId>
            <artifactId>hadoop-client</artifactId>
            <version>2.6.0-cdh5.14.2</version>
        </dependency>
    </dependencies>
< project>
EmpHCatMR.java
import org.apache.hadoop.conf.Configuration;
import org.apache.hadoop.io.IntWritable;
import org.apache.hadoop.io.Text;
import org.apache.hadoop.io.WritableComparable;
import org.apache.hadoop.mapreduce.Job;
import org.apache.hadoop.mapreduce.Mapper;
import org.apache.hadoop.mapreduce.Reducer;
import org.apache.hive.hcatalog.data.DefaultHCatRecord;
import org.apache.hive.hcatalog.data.HCatRecord;
import org.apache.hive.hcatalog.data.schema.HCatSchema;
import org.apache.hive.hcatalog.mapreduce.*;
import java.io.IOException;
public class EmpHCatMR {
    public static class Map extends
            Mapper<WritableComparable,HCatRecord,Text,IntWritable>{
        @Override
        protected void map(WritableComparable key,
                           HCatRecord value,
                           Context context)
                throws IOException, InterruptedException {
            //sex_age 字段格式： "[Male, 30]"
            Object sex_age=value.get("sex_age",
               HCatInputFormat.getTableSchema(context.getConfiguration()));
            String age=sex_age.toString().split(",")[1].replace("]","").trim();
            context.write(new Text("average-age"),
                        new IntWritable(Integer.parseInt(age)));
        }
    }
    //省略 Reduce 处理与 main()方法
}
```

HCatOutputFormat 使用 MapReduce 作业将处理后的数据写入 HCatalog 管理表中。所使用的主要方法介绍如下。

- setOutput(Job job,OutputJobInfo info)：设置作业输出信息，须指定数据库名和要写入的表名及分区，对应 OutputJobInfo 的三个参数为“databaseName”“tableName”“partitionValues”（Map<分区键，分区值>）。
- setSchema(Job job,HCatSchema schema)：设置最终写入数据的表结构信息。
- getTableSchema()：获得 HCatOutputFormat.setOutput()指定的表的结构信息。

示例 3-14

使用 HCatOutputFormat 将平均年龄统计结果写入 output 表中。

关键代码：

```
--创建 output 表
hive> use empdb;
hive> create table output(col1 string,col2 float);
EmpHCatMR.java
public static class Reduce extends
   Reducer<Text, IntWritable,WritableComparable, HCatRecord> {
         @Override
         protected void reduce(Text key, Iterable<IntWritable> values, Context context)
                                             throws IOException, InterruptedException {
               int sum = 0,i=0;
               for(IntWritable val:values){
                     sum+=val.get();
                     i++;
               }
               HCatRecord record = new DefaultHCatRecord(2);
               record.set(0, key);
               if(i>0) {
                     float avgAge=sum/i;
                     record.set(1, avgAge);
               }
               context.write(null, record);
         }
   }
   public static void main(String[] args) throws IOException,
               ClassNotFoundException,InterruptedException {
         String serverUri = "thrift://192.168.9.80:9083";
         String inputTableName = "emp";
         String outputTableName = "output";
         String dbName = "empdb";
         Configuration conf=new Configuration();
         conf.set("hive.metastore.uris",serverUri);
         Job job = new Job(conf, "Avg Age");
          // 初始化 HCatInputFormat
         HCatInputFormat.setInput(job,dbName,inputTableName);
         job.setInputFormatClass(HCatInputFormat.class);
```

```
        job.setJarByClass(EmpHCatMR.class);
        job.setMapperClass(Map.class);
        job.setReducerClass(Reduce.class);
        job.setMapOutputKeyClass(Text.class);
        job.setMapOutputValueClass(IntWritable.class);
        job.setOutputKeyClass(WritableComparable.class);
        job.setOutputValueClass(DefaultHCatRecord.class);
         // 初始化 HCatOutputFormat
        HCatOutputFormat.setOutput(job,
                        OutputJobInfo.create(dbName, outputTableName, null));
        HCatSchema s = HCatOutputFormat.getTableSchema(job.getConfiguration());
        HCatOutputFormat.setSchema(job, s);
        job.setOutputFormatClass(HCatOutputFormat.class);
        job.waitForCompletion(true);
    }
```

编译项目并运行，查看 output 表结果，如图 3.9 所示。

```
hive> select * from output;
OK
average-age     37.0
Time taken: 0.064 seconds, Fetched: 1 row(s)
```

图3.9　HCatalog MapReduce API示例项目运行结果

本章小结

➢ Hive 元数据是描述数据的数据，包括 hive 表的数据库名、表名、字段名称与类型、分区字段与类型等。

➢ Hive 元数据的存储方式包括单用户、多用户及远程模式 3 种。

➢ 元数据具体细分为按数据库、表、字段、分区等主要数据单元及相关信息存储。

➢ 使用 hive-metastore 提供的 HiveMetaStoreClient 类可对元数据进行编程管理。

➢ HCatalog 在元数据存储底层须对 HiveMetaStoreClient 进行调用。

➢ HCatalog CLI 兼容大部分 Hive DDL，用于对元数据进行管理。

➢ HCatalog 为不同大数据处理工具提供了元数据统一共享解决方案。

本章作业

一、简答题

1. 什么是 Hive 元数据？

2. Hive 有几种存储模式？它们分别是什么？

二、编码题

1. 使用 HCatalog CLI 创建零售商店顾客表（retail_db.customers），表结构如图 3.10 所示。

名	类型	长度
customer_id	int	11
customer_fname	varchar	45
customer_lname	varchar	45
customer_email	varchar	45
customer_password	varchar	45
customer_street	varchar	255
customer_city	varchar	45
customer_state	varchar	45
customer_zipcode	varchar	45

图3.10　零售商店顾客表结构

2．使用 HiveMetaStoreClient 获取零售商店顾客表的所有字段。

第 4 章

Hive 高级操作

技能目标

- 掌握 Hive 数据关联操作。
- 掌握 Hive 数据排序。
- 掌握 Hive 聚合操作。
- 掌握 Hive 窗口函数。

本章任务

任务 1　关联查询零售商店订单明细。
任务 2　使用分组排序实现商品销售量排行。
任务 3　使用窗口函数实现零售数据统计。

本章资源下载

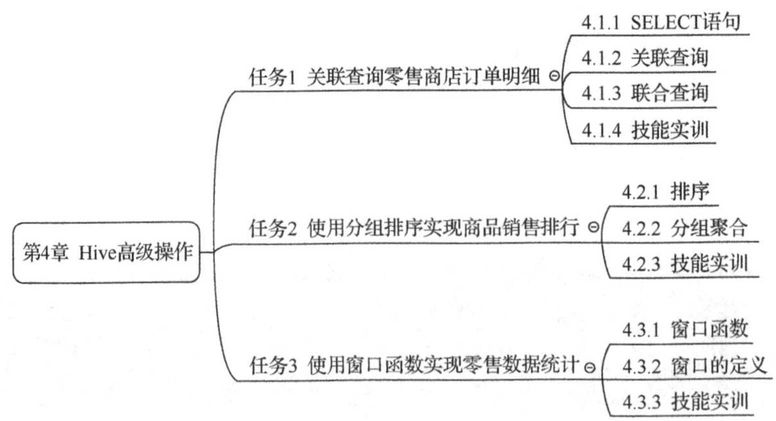

Hive SELECT 语句用于从表中检索数据，是标准 SQL 的子集，在 Hive 中也是使用频率最高、最复杂的部分。本章从基本的 SELECT 语句出发，逐渐深入讲解 SELECT 语句的列匹配正则表达式（REGEX Column Specification）、虚拟列（Virtual Columns）、关联查询（Join）、集合操作（Union）、排序、分组聚合以及窗口函数（Windowing）。

任务 1 关联查询零售商店订单明细

【任务描述】

零售商店业务数据（retail_db）包括顾客表（customers）、部门表（departments）、商品分类表（categories）、商品表（products）、订单表（orders）及订单明细表（order_items）共 6 张表。通过 Hive 查询出每个顾客购买的商品情况，包括商品名称、商品单价及订单总额。

【关键步骤】

（1）探索顾客数据。

（2）探索商品数据。

（3）探索订单及订单明细数据。

（4）关联查询顾客、商品、订单及订单明细表。

4.1.1 SELECT 语句

Hive SELECT 语句用于对表进行查询，即按照规定的语法规则从表中选取数据，并将查询结果保存在结果表中。其基本语法结构如下。

【语法】

```
SELECT [ALL | DISTINCT] select_expr,select,_expr,...
      FROM table_reference
       [WHERE where_condition]
       [GROUP BY col_list]
       [ORDER BY col_list]
```

[CLUSTER BY col_list | [DISTRIBUTE BY col_list] [SORT BY col_list]]
[LIMIT [offset,] rows]

1. where 子句

where 条件必须是布尔表达式，用于过滤结果集。

示例 4-1

查询所在州为“NY”，所在城市为“New York”的用户。

```
hive> select * from customers where customer_state='NY' and customer_city='New York';
```

查询结果如图 4.1 所示。

```
hive> select * from customers where customer_state='NY' and customer_city='New York';
OK
323	Joshua	Smith	XXXXXXXXX	XXXXXXXXX	3758 Sunny Mall	New York	NY	10025
336	Ruth	Armstrong	XXXXXXXXX	XXXXXXXXX	6146 Dewy Jetty New York	NY	10029
417	Teresa	Smith	XXXXXXXXX	XXXXXXXXX	1690 Dusty Mount	New York	NY	10024
585	Mary	Smith	XXXXXXXXX	XXXXXXXXX	3139 Golden Autumn Byway	New York	NY	10021
610	Mary	Smith	XXXXXXXXX	XXXXXXXXX	6759 Foggy Limits	New York	NY	10033
753	Sandra	Stafford	XXXXXXXXX	XXXXXXXXX	7557 Wishing Falls	New York	NY	10033
827	Tyler	Smith	XXXXXXXXX	XXXXXXXXX	5505 Grand Cloud Farm	New York	NY	10031
```

图4.1 where子句查询结果

Hive 支持的常用条件表达式的用法介绍如表 4-1 所示。

表 4-1 Hive 支持的常用条件表达式用法介绍

操作	操作数类型	描述
A=B	基本类型	A 与 B 相等则返回 TRUE，否则返回 FALSE
A==B	基本类型	同 A=B
A<=>B	基本类型	如果 A、B 均非空（NULL），同 A=B； 如果 A、B 均为空，返回 TRUE； 如果 A、B 有一个为空，返回 FALSE
A<>B	基本类型	A、B 均不为空时，若 A 不等于 B 则返回 TRUE，否则返回 FALSE； 如果 A、B 有一个为空，则返回 NULL
A!=B	基本类型	同 A<>B
A<B	基本类型	A、B 均不为空时，若 A 小于 B 则返回 TRUE，否则返回 FALSE； 如果 A、B 有一个为空，则返回 NULL
A<=B	基本类型	A、B 均不为空时，若 A 小于等于 B 则返回 TRUE，否则返回 FALSE； 如果 A、B 有一个为空，则返回 NULL
A>B	基本类型	A、B 均不为空时，若 A 大于 B 则返回 TRUE，否则返回 FALSE； 如果 A、B 有一个为空，则返回 NULL
A>=B	基本类型	A、B 均不为空时，若 A 大于等于 B 则返回 TRUE，否则返回 FALSE； 如果 A、B 有一个为空，则返回 NULL
A [NOT] BETWEEN B AND C	基本类型	A、B、C 均不为空时，若 A 大于等于 B 且 A 小于等于 C 则返回 TRUE，否则返回 FALSE，可以使用 NOT 进行反转 如果 A、B、C 有一个为空，则返回 NULL
A IS [NOT] NULL	所有类型	A 为空时，返回 TRUE，否则返回 FALSE，可使用 NOT 进行反转

续表

操作	操作数类型	描述
A [NOT] LIKE B	字符串类型	'abc' like 'a*'， 'abc' like 'a%', 'abc' like 'a__'（两个下划线） 上面三种用法均会返回 TRUE，可使用 NOT 进行反转； 如果 A、B 有一个为空，则返回 NULL
A RLIKE B	字符串类型	使用正则表达式匹配，例如： 'foobar' rlike '^f.*r$' 返回 TRUE; 如果 A、B 有一个为空，则返回 NULL
A REGEXP B	字符串类型	同 A RLIKE B
A AND B	布尔类型	A、B 均为 TRUE 则返回 TRUE，否则返回 FALSE
A OR B	布尔类型	A、B 有一个为 TRUE 则返回 TRUE，否则返回 FALSE
NOT A	布尔类型	A 为 FALSE，则返回 TRUE，否则返回 FALSE
!A	布尔类型	同 NOT A
A [NOT] IN (val1,val2,...)	基本类型	A 若出现在“值集合”中则返回 TRUE，若使用 NOT 反转表示未出现则返回 FALSE

注意

Hive where 子句中不允许出现子查询，即“where in (subquery)”暂不被支持，例如下列 HQL 语句将无法执行。

```
select * from customers where customer_id in
(select customer_id from customers where customer_state='NY')
```

2. ALL、DISTINCT 子句

ALL 与 DISTINCT 选项表示是否返回重复行，默认是 ALL，即返回所有匹配的行。大多数情况下使用 DISTINCT 子句指出在结果集中重复出现的行。

示例 4-2

查询出订单表中共有多少不同顾客下过订单。

分析：

订单表中存在一个顾客多次下单的情况，为满足查询需求，可使用 DISTINCT 对多次下单的每个顾客只保留他们各自的一条下单信息。

关键代码：

```
hive> select distinct order_customer_id from orders;
```

与不使用 DISTINCT 的结果对比如图 4.2 所示。

```
hive> select  order_customer_id from  orders;
Time taken: 0.048 seconds, Fetched: 68883 row(s)
hive> select distinct order_customer_id from  orders;
Time taken: 57.787 seconds, Fetched: 12405 row(s)
```

图4.2　DISTINCT子句使用与否结果对比

3. LIMIT 子句

LIMIT 子句用于限制 SELECT 语句返回的行数，其后的整型参数表示共返回多少行。

注意

从 2.0 版本开始 Hive 支持两个参数（均为整数），分别表示从指定位置（第一个参数）开始和返回指定的行数（第二个参数）。

示例 4-3

返回商品列中的前 5 个商品。

关键代码：

```
hive> select product_id,product_name from products limit 5;
```

输出结果如图 4.3 所示。

```
hive> select product_id,product_name from products limit 5;
OK
1       Quest Q64 10 FT. x 10 FT. Slant Leg Instant U
2       Under Armour Men's Highlight MC Football Clea
3       Under Armour Men's Renegade D Mid Football Cl
4       Under Armour Men's Renegade D Mid Football Cl
5       Riddell Youth Revolution Speed Custom Footbal
Time taken: 0.052 seconds, Fetched: 5 row(s)
```

图4.3　应用LIMIT子句的输出结果

4. 公共表表达式

公共表表达式（Common Table Expressions，CTE）可以表示一个临时的结果集（表），该表通过一个简单的查询指定，只要在 CTE 语句范围内均可共享该临时表。

【语法】

```
WITH t1 AS (SELECT ...) SELECT * FROM T1;
```

示例 4-4

从顾客表中查询每位顾客的全名。

关键代码：

```
hive> with t1 as (
    > select concat(customer_fname,'-',customer_lname) as fullname from customers
    > ) select * from t1 ;
```

上面用到了 concat 内置函数，其表示将多个字符串拼接为一个字符串。CTE 的范围在一条语句之间，注意“;”的位置。输出结果片段如图 4.4 所示。

```
Mary-Smith
Hannah-Brown
Mary-Rios
Angela-Smith
Benjamin-Garcia
Mary-Mills
Laura-Horton
Time taken: 0.042 seconds, Fetched: 12435 row(s)
```

图4.4　CET使用示例输出结果片段

5. 嵌套查询

嵌套查询也称为子查询，通常用于 FROM 子句后。

【语法】

SELECT … FROM (subquery) [AS] name …

- 子查询必须给定名称，因为 FROM 子句中的每个表必须有表名。
- 子查询中的列必须有唯一的名称，并且在外部查询中可以引用。
- 子查询中可以进行 UNION 和 JOIN 操作。
- Hive 支持任意级别的子查询。
- “AS”关键字在 Hive0.13 版本后才被支持。

示例 4-5

使用子查询从顾客表中查询每位顾客的全名。

关键代码：

```
hive> select * from (
    > select concat(customer_fname,'-',customer_lname) as fullname from customers) t1
    > limit 5;
```

输出结果如图 4.5 所示。

```
hive> select * from
    > (select concat(customer_fname,'-',customer_lname) as fullname from customers)
    > t1 limit 5;
OK
Richard-Hernandez
Mary-Barrett
Ann-Smith
Mary-Jones
Robert-Hudson
Time taken: 0.048 seconds, Fetched: 5 row(s)
```

图4.5　子查询使用示例输出结果

6. 列匹配正则表达式

Hive SELECT 语句支持使用正则表达式指定列名称，凡是符合正则表达式规则的列名将被作为结果集中的一列。

【语法】

SELECT `regex_expr` FROM table_reference

正则表达式须用一对反引号“`”引住，同时设置“hive.support.quoted.identifiers”属性为“none”，使 Hive 将反引号解释为正则表达式。

示例 4-6

使用正则表达式匹配顾客表中顾客 ID、姓名与所在城市列。

关键代码：

```
hive> set hive.support.quoted.identifiers=none;
    > select `[^_]*_(id|fname|lname|city)+` from customers limt 5;
```

输出结果如图 4.6 所示。

```
hive> select `[^_]*_(id|fname|lname|city)+` from customers limit 5;
OK
1       Richard Hernandez       Brownsville
2       Mary    Barrett Littleton
3       Ann     Smith   Caguas
4       Mary    Jones   San Marcos
5       Robert  Hudson  Caguas
Time taken: 0.046 seconds, Fetched: 5 row(s)
```

图4.6　列匹配正则表达式使用示例输出结果

关于正则表达式的具体语法规则请扫描二维码查看。

正则表达式的具体语法规则

7. *虚拟列*

虚拟列是并未在表中真正存在的列，但对于数据进行相关验证时非常有用。Hive 的两个常用虚拟列介绍如下。

➢ INPUT__FILE__NAME：包含 Mapper 任务运行时的输入文件名，也就是该行数据包含于哪个文件中。

➢ BLOCK__OFFSET__INSIDE__FILE：包含文件中的块内偏移量。

其中“__”为两个下划线。

示例 4-7

查看顾客表的虚拟列。

关键代码：

```
hive> select customer_id,input__file__name,block__offset__inside__file from customers;
```

输出结果如图 4.7 所示。

```
hive> select customer_id,input__file__name,block__offset__inside__file from customers limit 5;
OK
1       hdfs://cluster1/data/retail_db/customers/customers.csv  0
2       hdfs://cluster1/data/retail_db/customers/customers.csv  98
3       hdfs://cluster1/data/retail_db/customers/customers.csv  194
4       hdfs://cluster1/data/retail_db/customers/customers.csv  283
5       hdfs://cluster1/data/retail_db/customers/customers.csv  373
Time taken: 0.057 seconds, Fetched: 5 row(s)
```

图4.7　虚拟列使用示例输出结果

4.1.2　关联查询

关联查询是指对多表进行联合查询，主要通过 JOIN 语句将两个或多个表中的行组合在一起进行查询。HQL JOIN 类似于 SQL JOIN，但其仅支持等值连接，具体包括内连接（INNER JOIN）、外连接（OUTER JOIN）与交叉连接（CROSS JOIN）。

1. 内连接

内连接用于按连接条件组合两个表中的记录，返回两个表的交集。语法如下。

【语法】

```
table_reference [INNER] JOIN table_factor [join_condition]
```

其中，INNER 关键字通常会被省略，也就是说 JOIN 是 INNER JOIN 的简写。

示例 4-8

获取每个订单对应的购买商品的列表。

关键代码：

```
hive> select order_items.order_item.order_id,
    > order_items.order_item_product_id,
    > order_items.order_item_quantity,
    > order_items.order_item_subtotal,
    > order_items.order_item_product_price
    > from
    > orders join order_items on orders.order_id=order_items.order_item_order_id
    > limit 10;
```

这里只输出前 10 个连接结果，如图 4.8 所示。

```
OK
1       957     1       299.98  299.98
2       1073    1       199.99  199.99
2       502     5       250     50
2       403     1       129.99  129.99
4       897     2       49.98   24.99
4       365     5       299.95  59.99
4       502     3       150     50
4       1014    4       199.92  49.98
5       957     1       299.98  299.98
5       365     5       299.95  59.99
```

图4.8　JOIN内连接使用示例输出结果

从图 4.8 中可以看出，订单 2 中共有 3 件商品，编号 502 商品共购买 5 件，总价 250。可使用别名对上述代码进行简化，简化后的代码如下。

```
hive> select b.order_item.order_id,
    > b.order_item_product_id,
    > b.order_item_quantity,
    > b.order_item_subtotal,
    > b.order_item_product_price
    > from
    > orders a join order_items b on a.order_id=b.order_item_order_id
    > limit 10;
```

2. 外连接

外连接分为 3 类：左外连接（LEFT OUTER JOIN）、右外连接（RIGHT OUTER JOIN）和全外连接（FULL OUTER JOIN）。

（1）左外连接

左外连接简称左连接。由内连接可以看出，结果集中的记录是在连接的两个表中都存在的记录。左连接的区别是左表的记录全部被选择，而右表只选择符合连接条件的记录，最后将无法与左表中记录进行对应的右表记录显示为 NULL。

【语法】

table_reference LEFT [OUTER] JOIN table_factor [join_condition]

示例 4-9

使用左连接获取没下订单的所有顾客信息。

关键代码：

```
hive> select c.customer_id,o.order_id
    > from
    > customers c left join orders o on c.customer_id=o.order_customer_id
    > where o.order_id is null limit 20;
```

采用左连接后，所有顾客编写都将出现在结果集中，而与左边顾客无法对应的右边订单将显示为 NULL，这里仅输出结果集前 20 条记录，如图 4.9 所示。

```
219     NULL
339     NULL
469     NULL
1187    NULL
1481    NULL
1808    NULL
2073    NULL
2096    NULL
2450    NULL
4555    NULL
4927    NULL
6072    NULL
6613    NULL
7011    NULL
7552    NULL
8243    NULL
8343    NULL
8575    NULL
8778    NULL
8882    NULL
```

图4.9 左连接使用示例输出结果

从图中可看出，其中编号为 219 的顾客没有产生任何订单。

（2）右外连接

右外连接简称右连接。与左连接相反，右连接将右表的记录全部选择，而将左表中没有符合连接条件的记录显示为 NULL。

【语法】

table_reference RIGHT [OUTER] JOIN table_factor [join_condition]

示例 4-10

使用右连接获取没下订单的所有顾客信息。

分析：

与示例 4-9 的需求相同，将示例 4-9 的代码换为右连接，然后调换表的左右顺序即可。

关键代码：

```
hive> select c.customer_id,o.order_id
```

> from
> orders o right join customers c on o.order_customer_id=c.customer_id
> where o.order_id is null limit 10;

输出结果与示例 4-9 相同。

（3）全外连接

全外连接结合了左连接和右连接的结果，即：对左表而言，右表中不符合连接条件的记录显示为 NULL；对右表而言，左表中不符合连接条件的记录显示为 NULL。语法如下。

【语法】

table_reference FULL [OUTER] JOIN table_factor [join_condition]

全外连接示意如图 4.10 所示。

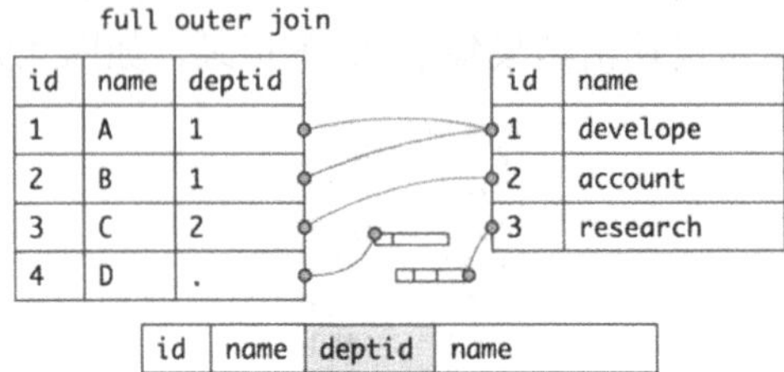

id	name	deptid	name
1	A	1	develope
2	B	1	develope
3	C	2	account
4	D		
		3	research

图4.10　全外连接

3. 交叉连接

交叉连接又称笛卡尔乘积，相当于两个表相乘。比如表 A 交叉连接表 B，则表 A 中的每一条记录都与表 B 中的每一条记录连接。语法如下。

【语法】

table_reference CROSS JOIN table_factor [join_condition]

交叉连接示意如图 4.11 所示。

R

A	B	C
a_1	b_1	c_1
a_1	b_2	c_2
a_2	b_2	c_1

S

A	B	C
a_1	b_2	c_2
a_1	b_3	c_2
a_2	b_2	c_1

R×S

R.A	R.B	R.C	S.A	S.B	S.C
a_1	b_1	c_1	a_1	b_2	c_2
a_1	b_1	c_1	a_1	b_3	c_2
a_1	b_1	c_1	a_2	b_2	c_1
a_1	b_2	c_2	a_1	b_2	c_2
a_1	b_2	c_2	a_1	b_3	c_2
a_1	b_2	c_2	a_2	b_2	c_1
a_2	b_2	c_1	a_1	b_2	c_2
a_2	b_2	c_1	a_1	b_3	c_2
a_2	b_2	c_1	a_2	b_2	c_1

图4.11　交叉连接

4.1.3　联合查询

UNION 语句用于合并两个或多个 SELECT 语句的结果集。

【语法】

```
select_statement UNION [ALL|DISTINCT]
select_statement UNION [ALL|DISTINCT] select_statement ...
```

➢ UNION 的每个子集都必须有相同的列名和类型。

➢ 应在获得整个 UNION 结果之后进行排序、分组、LIMIT 等操作。

注意

1.2 版本前的 Hive 只支持 UNION ALL，重复的行不会被删除。

示例 4-11

获取分类编号分别为 2 和 3 的商品列表。

关键代码：

```
hive> select product_id,product_category_id,product_name
    > from products where product_category=2
    > union all
    > select product_id,product_category_id,product_name
    > from products where product_category=3
```

部分输出结果如图 4.12 所示。

```
19	2	Nike Men's Fingertrap Max Training Shoe
20	2	Under Armour Men's Highlight MC Football Clea
21	2	Under Armour Kids' Highlight RM Football Clea
22	2	Kijaro Dual Lock Chair
23	2	Under Armour Men's Highlight MC Alter Ego Hul
24	2	Elevation Training Mask 2.0
25	3	Quest Q64 10 FT. x 10 FT. Slant Leg Instant U
26	3	Nike Men's USA White Home Stadium Soccer Jers
27	3	Nike Youth USA Away Stadium Replica Soccer Je
```

图4.12　UNION联合查询示例的部分输出结果

4.1.4　技能实训

请根据零售商店业务数据查询所有订单的明细，并将结果集保存为新的订单表 order_details。新表字段包含内容如下。

➢ 订单：编号、日期。

➢ 订单顾客：姓名、所在城市。

➢ 订单商品：名称、分类名称、单价、购买数量。

输出结果如表 4-2 所示。

表 4-2　订单联合查询输出结果

订单 ID	顾客	城市	商品名	商品分类	日期	单价	数量
2	Rodriguez	Chicago	Nike ...	Wo...	2013-07-25	50	5

分析：

- 从订单表中可获取订单 ID 与日期。
- 从顾客表中可获取顾客姓名与所在城市，但须连接订单表与顾客表。
- 从订单明细表中可获取商品的编号、分类编号、单价及数量，但须连接订单明细表、商品表与分类表以获取商品的名称和分类名称。

任务 2　使用分组排序实现商品销售排行

【任务描述】

通过 HQL 分组排序功能实现订单中每个分类下商品销售排行。

【关键步骤】

（1）对订单中的商品销售数量进行（降序）排序。

（2）按商品分类进行分组。

（3）统计组内商品销量。

4.2.1　排序

Hive 提供了 4 种排序方式：ORDER BY、SORT BY、DISTRIBUTE BY 及 CLUSTER BY。

1. ORDER BY

Hive 中的 ORDER BY 语句与 SQL 中的类似，可以对结果集进行全局排序，即 Hive 可对所有数据进行 Reducer 处理以保证全局有序，但当数据规模较大时此过程比较耗时。所以在 strict（hive.mapred.mode=strict，默认是 nonstrict）模式下，Hive 对 ORDER BY 进行了限制，要求 ORDER BY 子句后必须跟随“LIMIT”子句，以防止单个 Reducer 处理时间过长。

【语法】

```
ORDER BY colName (ASC|DESC)? (,colName (ASC|DESC)?)*
```

ORDER BY 语句默认按 ASC（升序）排序，排序字段必须出现在 SELECT 子句中。

ORDER BY 性能较差，应尽量在排序前对数据进行过滤。

示例 4-12

统计 order_items 表中销量最多的商品。

关键代码：

```
hive> set hive.mapred.mode=strict;
    > select order_item_order_id,order_item_product_id,order_item_quantity
    > from order_items order by order_item_quantity desc limit 10;
```

输出结果如图 4.13 所示。

order	product	quantity
52599	1014	5
52599	1014	5
52598	1014	5
45210	627	5
67056	1014	5
45213	365	5
5272	792	5
52595	572	5
45215	1014	5
45215	502	5

图4.13 销量最多的商品

从图 4.13 中可以看出，编号为 1014 的商品被购买数量最多。

提示

使用 hive.cli.print.header=true 可以打印查询结果的表头信息。

有些情况下，我们不清楚需要按哪一列进行排序，或者说这个待排序列是动态的，例如：

```
select (case when a is null then b else a end) sort_col from table order by sort_col
```

在这种情况下，当表中 a 列为空时按 b 列排序，否则仍按 a 列排序。ORDER BY 后面支持使用 CASE WHEN 语句，等价于上面写法。例如：

```
select a,b from table order by case when a is null then b else a end
```

2. SORT BY

SORT BY 指出了数据在每个 Reducer 内如何排序，即 SORT BY 只在每个 Reducer 内对数据进行排序。语法上 SORT BY 与 ORDER BY 相似，但从作用上看，ORDER BY 可以看作是 SORT BY 的一个特例：当 SORT BY 的 Reducer 数目设置为 1 时，等同于 ORDER BY。所以两者真正的区别是：ORDER BY 强调全局有序，SORT BY 只强调 Reducer 内局部有序。

示例 4-13

查看商品分类表 categories，结果按分类名称 category_name 升序排列。

关键代码：

```
hive> set mapred.reduce.tasks=2;
    > select * from categories sort by category_name;
```

部分输出结果如图 4.14 所示。

```
18      4       Men's Footwear
30      6       Men's Golf Clubs
8       2       More Sports
52      8       NBA
50      8       NFL
58      8       NFL Players
29      5       Shop By Sport
10      3       Strength Training
39      6       Team Shop
28      5       Top Brands
41      6       Trade-In
24      5       Women's Apparel
------------------------------------------
40      6       Accessories
22      4       Accessories
25      5       Boys' Apparel
43      7       Camping & Hiking
9       3       Cardio Equipment
45      7       Fishing
26      5       Girls' Apparel
32      6       Golf Apparel
34      6       Golf Bags & Carts
36      6       Golf Balls
33      6       Golf Shoes
7       2       Hockey
44      7       Hunting & Shooting
```

图4.14　SORT BY使用示例部分输出结果

从图 4.14 中可以看出，分隔线以上部分为第一个 Reducer 的排序结果，分隔线以下部分则是第二个 Reducer 的排序结果。通常 SORT BY 不会单独使用，而是同 DISTRIBUTE BY 一起使用。

3. DISTRIBUTE BY

DISTRIBUTE BY 控制 Map 的输出数据在 Reducer 中的划分。如果 DISTRIBUTE BY 指定列的值相同，则它们会被送到同一个 Reducer 中进行处理。

注意

DISTRIBUTE BY 必须写在 SORT BY 之前。

示例 4-14

查看商品分类表 categories，结果按分类名称 category_name 升序排列，要求同一大类（即 category_department_id 相同）的商品分类（category_id）在同一个 Reducer 中处理。

关键代码：

```
hive> set mapred.reduce.tasks=2;
    > select * from categories distribute by category_department_id sort by category_name;
```

部分输出结果如图 4.15 所示。

53	8	NCAA
50	8	NFL
58	8	NFL Players
51	8	NHL
2	2	Soccer
39	6	Team Shop
6	2	Tennis & Racquet
41	6	Trade-In
19	4	Women's Footwear
31	6	Women's Golf Clubs
56	8	World Cup Shop
27	5	Accessories
16	3	As Seen on TV!
42	7	Bike & Skate Shop
47	7	Boating
12	3	Boxing & MMA
25	5	Boys' Apparel
43	7	Camping & Hiking
9	3	Cardio Equipment
13	3	Electronics
45	7	Fishing
11	3	Fitness Accessories
26	5	Girls' Apparel
44	7	Hunting & Shooting
46	7	Indoor/Outdoor Games

图4.15　DISTRIBUTE BY使用示例部分输出结果

从图 4.15 中可以看出，分隔线以上部分为第一个 Reducer 的排序结果，第 2 列 category_department_id 为偶数的商品信息在该 Reducer 中排序；而分隔线以下部分则是第二个 Reducer 的排序结果，第 2 列 category_department_id 为奇数的商品信息在该 Reducer 中排序。

4. CLUSTER BY

CLUSTER BY 相当于 DISTRIBUTE BY 和 SORT BY 的结合体。当 DISTRIBUTE BY 和 SORT BY 都在同一列上操作时，如下 2 个语句是等价的。

```
select A,B,C from table distribute by A sort by A
select A,B,C from table cluster by A
```

另外，CLUSTER BY 指定的列只能进行升序排序，无法手动指定排序方向。

Hive 各种排序操作的演示与对比请扫描二维码查看。

4.2.2 分组聚合

Hive 排序操作的演示与对比

Hive 提供了多种聚合函数。所谓聚合函数是指对一组值进行计算并返回单个值的函数，其通常与 SELECT 语句的 GROUP BY 子句一起使用。

1. **分组**

SELECT 语句在未指定 GROUP BY 子句的情况下，会将整个表当作一个分组。例如 count()函数便是对组内的所有行数进行统计，其默认统计整个表的行数而不需要指定 GROUP BY。

又如在示例 4-12 中，仍无法获知销量最多的商品，因为大部分商品均是以多行记录出现。现在必须进行分组合并，即将相同的商品归为一组，然后在组内进行统计。

【语法】

```
SELECT expression (,expression)* FROM src
GROUP BY expression (,expression)* HAVING condition
```

- 除了聚合函数外，SELECT 所选择的列也必须出现在 GROUP BY 子句中。
- GROUP BY 支持使用 CASE WHEN 表达式。
- GROUP BY 配合使用 HAVING 进行过滤。

示例 4-15

统计 order_items 表中销量排名前 10 的商品。

关键代码：

```
hive> select order_item_product_id,sum(order_item_quantity) as total
    > from order_items
    > group by order_item_product_id
    > order by total desc limit 10;
```

输出结果如图 4.16 所示。

```
OK
order_item_product_id   total
365     73698.0
502     62956.0
1014    57803.0
191     36680.0
627     31735.0
403     22246.0
1004    17325.0
1073    15500.0
957     13729.0
977     998.0
Time taken: 145.677 seconds, Fetched: 10 row(s)
```

图4.16 商品销量Top 10

上述代码用到了聚合函数 sum()，其表示对组内的某列进行求和。如果须对分组聚合后的结果进行过滤，可以使用 HAVING。例如，统计商品销量不超过 1000 的 Top 10，代码如下：

```
hive> select order_item_product_id,sum(order_item_quantity) as total
    > from order_items
    > group by order_item_product_id having total<=1000
    > order by total desc limit 10;
```

2. 基础聚合

常用的内置聚合函数列举如下。

- max(col)：返回组内某列中的最大值。
- min(col)：返回组内某列中的最小值。
- count(*)：返回组内总行数，包括值为 NULL 的行。
- count(expr)：返回组内 expr 表达式不是 NULL 的总行数。
- count(DISTINCT expr)：返回组内 expr 唯一且非 NULL 的行的数量。
- sum(col)：返回组内某列元素的总和。
- avg(col)：返回组内某列元素的平均值。

➢ collect_set(col)：返回消除了重复元组的数组。
➢ collect_list(col)：返回允许存在重复元素的数组。

示例 4-16

统计每个商品大类下的商品分类。

关键代码：

```
hive> select category_department_id,count(category_name),collect_set(category_name)
    > from categories group by category_department_id;
```

输出结果如图 4.17 所示。

```
OK
category_department_id  _c1     _c2
2       8       ["Football","Soccer","Baseball & Softball","Basketball","Lacrosse","Tennis & Racquet","Hockey","More Spor
ts"]
4       6       ["Cleats","Men's Footwear","Women's Footwear","Kids' Footwear","Featured Shops","Accessories"]
6       12      ["Men's Golf Clubs","Women's Golf Clubs","Golf Apparel","Golf Shoes","Golf Bags & Carts","Golf Gloves","G
olf Balls","Electronics","Kids' Golf Clubs","Team Shop","Accessories","Trade-In"]
8       10      ["MLB","NFL","NHL","NBA","NCAA","MLS","International Soccer","World Cup Shop","MLB Players","NFL Players"
]
3       8       ["Cardio Equipment","Strength Training","Fitness Accessories","Boxing & MMA","Electronics","Yoga & Pilate
s","Training by Sport","As Seen on  TV!"]
5       7       ["Men's Apparel","Women's Apparel","Boys' Apparel","Girls' Apparel","Accessories","Top Brands","Shop By S
port"]
7       7       ["Bike & Skate Shop","Camping & Hiking","Hunting & Shooting","Fishing","Indoor/Outdoor Games","Boating","
Water Sports"]
```

图4.17　每个商品大类下的商品分类

3. 高级聚合

在 Hive 中，新增了几种增强聚合功能。

（1）GROUPING SETS

GROUPING SETS 可以实现对同一个数据集进行多重 GROUP BY 操作，这本质上是多个 GROUP BY 进行 UNION ALL 操作。GROUPING SETS 与 GROUP BY 的等价操作介绍如表 4-3 所示。

表 4-3　GROUPING SETS 与 GROUP BY 的等价操作

GROUPING SETS	GROUP BY
select a,b,sum(c) from tab1 group by a,b grouping sets((a,b))	select a,b,sum(c) from tab1 group by a,b
select a,b,sum(c) from tab1 group by a,b grouping sets((a,b),a)	select a,b,sum(c) from tab1 group by a,b union all select a,null,sum(c) from tab1 group by a
select a,b,sum(c) from tab1 group by a,b grouping sets(a,b)	select a,null,sum(c) from tab1 group by a union all select null,b,sum(c) from tab1 group by b
select a,b,sum(c) from tab1 group by a,b grouping sets((a,b),a,b,())	select a,b,sum(c) from tab1 group by a,b union all select a,null,sum(c) from tab1 group by a,null union all select null,b,sum(c) from tab1 group by null,b union all select null,null,sum(c) from tab1

由表 4-3 可以看出 GROUPING SETS()中可包括多个分组元素，如果元素以“()”形式出现则被视为一个整体进行分组，分组元素间进行 UNION 操作。

（2）CUBE 与 ROLLUP

CUBE 与 ROLLUP 的功能比 GROUPING SETS 的功能强。CUBE 子句可对分组列进行所有可能的组合聚合。ROLLUP 子句用于在维度结构中计算聚合。

GROUP BY a,b,c WITH CUBE 等价于 GROUP BY a,b,c GROUPING SETS((a,b,c),(a,b),(a,c),(b,c),(a),(b),(c),())

GROUP BY a,b,c WITH ROLLUP 等价于 GROUP BY a,b,c GROUPING SETS((a,b,c),(a,b),,(a) ())

总的来说，Hive 的增强聚合功能方便了部分有规律代码的编写，或者说缩短了代码的长度，其本质仍然是 GROUP BY。

4.2.3 技能实训

请根据零售商店业务数据查询消费金额排名前 10 的顾客。

分析：

➢ 关联查询订单表与顾客表以获取顾客-订单对应关系；

➢ 关联查询订单表与订单明细表以获取订单-订单总金额对应关系；

➢ 根据顾客-订单-订单总金额的关系，对顾客进行分组，并在组内对消费总额进行求和、排序。

任务 3 使用窗口函数实现零售数据统计

【任务描述】

通过窗口函数实现零售数据统计。

【关键步骤】

（1）掌握窗口函数的使用。

（2）掌握窗口的定义。

4.3.1 窗口函数

自 0.11.0 版本开始，Hive 加入了窗口函数功能。窗口函数是一组特殊的函数，它能扫描多个输入行以计算各输出值，可为每行数据分别生成一行结果记录，几乎所有复杂的聚合计算都可以通过它来完成。

【语法】

```
function(arg1,...) OVER([PARTITION BY <...>] [ORDER BY <...> [<window_clause>]])
```

➢ 窗口函数会作为 SELECT 语句中的一列出现，类似基本聚合函数，如 count(*)。

➢ OVER()表示在当前查询的结果集上进行操作，操作有分区与排序两种，均可选。下文中将把 OVER 操作的数据集简称为 OVER 对象。

➢ PARTITION BY 类似于 GROUP BY，表示对当前结果集按其中某列进行分组。如果未指定该子句，则意味着整个 SELECT 结果集作为一个分组。

➢ 只有在指定 ORDER BY 子句后才能进行窗口定义（window_clause），窗口定义不常用，但其功能强大，下一小节将详细介绍窗口的定义。

➢ 在一个 SELECT 语句中，可以多次出现窗口函数，即可同时使用多个窗口函数。

➢ 如果要对窗口函数的计算结果进行过滤，必须在窗口函数所在的 SELECT 语句往外一层进行操作。

Hive 提供多种窗口函数，按功能不同可划分为 3 类：排序类、聚合类及分析类。

1. **排序类**

（1）ROW_NUMBER()

该函数基于 OVER 对象分组、排序的结果，为每一行分组记录返回一个序号。该序号从 1 开始递增，遇到新组则重新从 1 开始递增。也就是说，该函数计算的值表示每组内部排序后的顺序编号。

注意

ROW_NUMBER 产生的序号总是连续的。

示例 4-17

查询 categories 表，使用 ROW_NUMBER()函数对每个商品大类下的子类进行排名，排名依据是子类编号大小。

关键代码：

```
hive> select *,
    > row_number()
    > over(partition by category_department_id order by cast(category_id as int))
    > from categories;
```

上述代码用到了 cast()类型转换函数，代码运行结果如图 4.18 所示。

```
category_id     category_department_id  category_name   _wcol0
1       2       Football        1
2       2       Soccer  2
3       2       Baseball & Softball     3
4       2       Basketball      4
5       2       Lacrosse        5
6       2       Tennis & Racquet        6
7       2       Hockey  7
8       2       More Sports     8
9       3       Cardio Equipment        1
10      3       Strength Training       2
11      3       Fitness Accessories     3
```

图4.18　ROW_NUMBER示例代码运行结果

（2）RANK ()

该函数与 ROW_NUMBER 类似，但在产生序号的连续性上的规则不同。比如针对下列情况，即两名学生的成绩并列第一，那么 RANK 规则会将两名学生作为并列第一名，接下来的学生作为第三名，之所以如此是因为第一名出现 2 次，第二名被跳过了。

示例 4-18

使用 RANK()函数实现每个订单中商品购买数量排序（由高到低）。

关键代码：

```
hive> select order_item_order_id as order,
    > order_item_product_id as product,
    > order_item_quantity as num,
    > rank()
    > over(partition by order_item_order_id order by order_item_quantity desc) as rank
    > from order_items limit 10;
```

代码运行结果如图 4.19 所示。

```
order   product num     rank
1       957     1       1
10      1014    2       1
10      1073    1       2
10      403     1       2
10      917     1       2
10      1073    1       2
100     191     3       1
100     365     2       2
100     403     1       3
1000    1014    2       1
```

图4.19　RANK示例代码运行结果

（3）DENSE_RANK ()

使用 RANK()排序时，一旦出现相同序号，后续序号将不再连续。与 RANK()相比，DENSE_RANK()不会跳号，将保持序号连续。表 4-4 说明了在同一组内按“score”排序分别使用ROW_NUMBER()、RANK()及DENSE_RANK()的区别。

表 4-4　分别使用 ROW_NUMBER()、RANK()及 DENSE_RANK()进行排序的区别

score	row_number()	rank()	dense_rank()
99	1	1	1
99	2	1	1
98	3	3	2

（4）NTILE(n)

NTILE(n)将 OVER 对象的分组结果数据集按照顺序平均切分成 n 片，并为每一行记录返回一个切片号。

示例 4-19

查询每个客户最近的前 1/3 订单。

关键代码：

```
hive> select order_customer_id,order_id,order_date,
    > ntile(3)
    > over(partition by order_customer_id order by order_date desc)
    > from orders limit 10;
```

代码运行部分结果如图 4.20 所示。

图 4.20 中的框内 3 行数据是编号为“100”的客户最近前 1/3 的订单，应注意到在切片不均匀时会增加第一个切片的数据。

（5）PERCENT_RANK()

该函数返回 OVER 对象分组内当前行的 RANK 值与组内总行数的比值。设当前行的 RANK 值为 rank，分组内的总行数为 rows，则该函数的具体公式可表示为：(rank-1)/(rows-1)。

```
order_customer_id        order_id        order_date        _wcol0
1       22945   2013-12-13 00:00:00     1
10      56133   2014-07-15 00:00:00     1
10      45239   2014-05-01 00:00:00     2
100     54995   2014-07-08 00:00:00     1
100     64426   2014-04-06 00:00:00     1
100     62907   2014-02-06 00:00:00     1
100     28477   2014-01-16 00:00:00     2
100     22395   2013-12-09 00:00:00     2
100     15045   2013-10-28 00:00:00     3
100     6641    2013-09-05 00:00:00     3
```

图4.20　NTILE(n)切片示例代码运行部分结果

示例 4-20

求商品分类的 PERCENT_RANK 值。

关键代码：

```
hive> select *,
    > percent_rank()
    > over(partition by category_department_id order by cast(category_id as int))
    > from categories;
```

代码运行部分结果如图 4.21 所示。

```
category_id     category_department_id  category_name   _wcol0
1       2       Football        0.0
2       2       Soccer  0.14285714285714285
3       2       Baseball & Softball     0.2857142857142857
4       2       Basketball      0.42857142857142855
5       2       Lacrosse        0.5714285714285714
6       2       Tennis & Racquet        0.7142857142857143
7       2       Hockey  0.8571428571428571
8       2       More Sports     1.0
9       3       Cardio Equipment        0.0
10      3       Strength Training       0.14285714285714285
11      3       Fitness Accessories     0.2857142857142857
12      3       Boxing & MMA    0.42857142857142855
13      3       Electronics     0.5714285714285714
14      3       Yoga & Pilates  0.7142857142857143
15      3       Training by Sport       0.8571428571428571
16      3       As Seen on  TV! 1.0
17      4       Cleats  0.0
18      4       Men's Footwear  0.2
19      4       Women's Footwear        0.4
20      4       Kids' Footwear  0.6
```

图4.21　PERCENT_RANK示例代码运行部分结果

2. 聚合类

常用的聚合类包括 COUNT(col)、SUM(col)、MAX(col)、MIN(col)及 AVG(col)等，分别返回 OVER 对象分组内的总行数、总和、最大值、最小值及平均值。

示例 4-21

使用聚合窗口函数在订单明细表中统计各订单中不同商品总数、订单总金额、订单最高/最低/平均金额。

关键代码：

```
hive> select order_item_order_id as order,
    > count(*) over(partition by order_item_order_id) as row_cnt,
    > sum(order_item_subtotal) over(partition by order_item_order_id) as sum_col,
    > max(order_item_subtotal) over(partition by order_item_order_id) as max_col,
```

```
> min(order_item_subtotal) over(partition by order_item_order_id) as min_col,
> avg(order_item_subtotal) over(partition by order_item_order_id) as avg_col
> from order_items limit 10;
```

代码运行结果如图 4.22 所示。

```
order   row_count       sum_col max_col min_col avg_col
1       1       299.98  299.98  299.98  299.98
10      5       651.9200000000001       99.96   129.99  130.38400000000001
10      5       651.9200000000001       99.96   129.99  130.38400000000001
10      5       651.9200000000001       99.96   129.99  130.38400000000001
10      5       651.9200000000001       99.96   129.99  130.38400000000001
10      5       651.9200000000001       99.96   129.99  130.38400000000001
100     3       549.94  299.97  119.98  183.31333333333336
100     3       549.94  299.97  119.98  183.31333333333336
100     3       549.94  299.97  119.98  183.31333333333336
1000    3       279.93  99.96   129.99  93.31
```

图4.22　聚合窗口函数示例代码运行结果

3. 分析类

➢ CUME_DIST()：返回小于等于当前值的行数与分组内总行数的比值。

➢ LAG/LEAD (col,n,DEFAULT)：统计窗口内往上/下第 n 行的值。第一个参数为列名，第二参数为往上/下第 n 行（可选，默认为 1），第三个参数表示当往上/下第 n 行为 NULL 时取该默认值，默认为 NULL。

➢ FIRST_VALUE/LAST_VALUE(col)：返回 OVER 对象分组内第一个值/最后一个值。

示例 4-22

使用分析窗口函数对商品表进行以下统计。

➢ 所有商品中低于当前商品价格的商品比例。

➢ 同类商品中低于当前商品价格的商品比例。

➢ 同类商品中比当前商品价格低的第一个商品。

➢ 同类商品中比当前商品价格高的第一个商品。

➢ 同类商品中价格最低的商品。

➢ 同类商品中价格最高的商品。

关键代码：

```
hive> select product_id as product,product_category_id as category,product_price as price,
    > cume_dist() over(order by cast(product_price as float)) as c1,
    > cume_dist()
    > over(partition by product_category_id order by cast(product_price as float)) as c2,
    > lag(product_price,1)
    > over(partition by product_category_id order by cast(product_price as float)) as c3,
    > lead(product_price,1)
    > over(partition by product_category_id order by cast(product_price as float)) as c4,
    > first_value(product_price)
    > over(partition by product_category_id order by cast(product_price as float)) as c5,
    > first_value(product_price)
    > over(partition by product_category_id order by cast(product_price as float) desc) as c6
```

> from products limit 20;

代码运行结果如图 4.23 所示。

product	category	price	c1	c2	c3	c4	c5	c6
208	10	1999.99	1.0	1.0	1799.99	NULL	34.99	1999.99
199	10	1799.99	0.9992565055762082	0.9583333333333334	999.99	1999.99	34.99	1999.99
197	10	999.99	0.9962825278810409	0.9166666666666666	499.99	1799.99	34.99	1999.99
209	10	499.99	0.979182156133829	0.875	399.99	999.99	34.99	1999.99
203	10	399.99	0.9650557620817843	0.8333333333333334	349.98	499.99	34.99	1999.99
202	10	349.98	0.9338289962825279	0.7916666666666666	309.99	399.99	34.99	1999.99
200	10	309.99	0.9286245353159851	0.75	299.99	349.98	34.99	1999.99
196	10	299.99	0.9263940520446097	0.7083333333333334	299.99	309.99	34.99	1999.99
198	10	299.99	0.9263940520446097	0.7083333333333334	299.98	299.99	34.99	1999.99

图4.23　分析窗口函数示例代码运行结果

注意

在统计"同类商品中价格最高的商品"时并没有使用 LAST_VALUE()函数，而是使用了 FIRST_VALUE()函数并结合 product_price 倒序实现。原因是：对于 LAST_VALUE()函数而言，使用默认的窗口子句（窗口定义），代码运行结果可能会出乎意料。

将上述代码中的 c6 结果列修改为如下代码：

```
last_value(product_price)
over(partition by product_category_id order by cast(product_price as float) ) as c6
```

代码运行结果如图 4.24 所示。

product	category	price	c1	c2	c3	c4	c5	c6
212	10	34.99	0.2988847583643123	0.041666666666666664	NULL	79.99	34.99	34.99
201	10	79.99	0.4788104089219331	0.08333333333333333	34.99	99	34.99	79.99
194	10	99	0.5591078066914498	0.125	79.99	99.95	34.99	99
195	10	99.95	0.5702602230483271	0.16666666666666666	99	159.99	34.99	99.95
207	10	159.99	0.7895910780669145	0.25	99.95	159.99	34.99	159.99
204	10	159.99	0.7895910780669145	0.25	159.99	169.99	34.99	159.99
215	10	169.99	0.8022304832713755	0.2916666666666667	159.99	189	34.99	169.99
216	10	189	0.8215613382899628	0.3333333333333333	169.99	199.98	34.99	189
210	10	199.98	0.8312267657992565	0.375	189	199.99	34.99	199.98
211	10	199.99	0.857992565055762	0.4583333333333333	199.98	199.99	34.99	199.99
213	10	199.99	0.857992565055762	0.4583333333333333	199.99	269.99	34.99	199.99

图4.24　错误的LAST_VALUE示例代码运行结果

从图 4.24 中可以看出，由于默认是按价格升序，所以当前行的价格总是最高价格，也就是说当前行的值总是最后一个值。因为默认的窗口子句为：

```
RANGE BETWEEN UNBOUNDED PRECEDING AND CURRENT ROW
```

所以，为了完成"同类商品中价格最高的商品"的统计，有两种方式可进行变通：一种如示例 4-22 中的代码，另一种便是自行窗口定义。

更多 Hive 窗口函数应用请扫描二维码了解。

Hive 窗口函数应用

4.3.2　窗口的定义

窗口的定义由窗口子句"[<window_clause>]"来实现，用于进一步细分分组结果并应用分析函数。窗口子句不支持的函数包括：RANK、

NTILE、DENSE_RANK、CUME_DIST、PERCENT_RANK、LEAD 及 LAG。

窗口分为两类：行类型窗口（行窗口）与范围类型窗口（范围窗口）。

1. 行窗口

行窗口是根据当前行之前或之后的行号确定的窗口。

【语法】

ROWS BETWEEN start_expr AND end_expr

其中，start_expr 可设置为下列值。

- UNBOUNDED PRECEDING：窗口起始位置，为分组的第一行。
- CURRENT ROW：当前行。
- *n* PRECEDING/FOLLOWING：当前行之前/之后 *n* 行。

end_expr 可设置为下列值。

- UNBOUNDED FOLLOWING：窗口结束位置，为分组的最后一行。
- CURRENT ROW：当前行。
- *n* PRECEDING/FOLLOWING：当前行之前/之后 *n* 行。

行窗口参数具体介绍如图 4.25 所示。

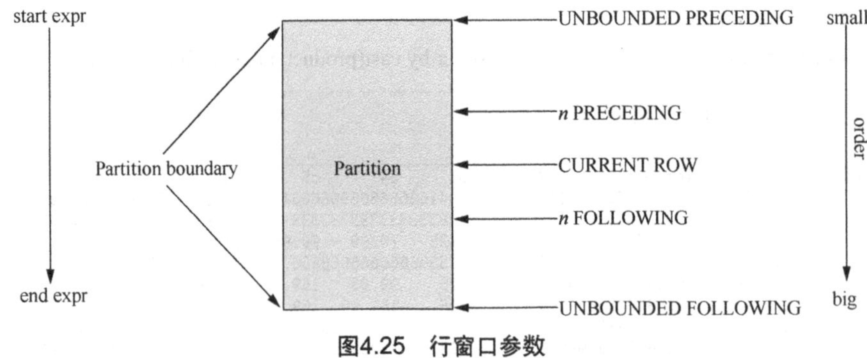

图4.25　行窗口参数

示例 4-23

使用 LAST_VALUE()统计同类商品中价格最高的商品。

关键代码：

```
hive> select product_id as product,product_category_id as category,product_price as price,
    > last_value(product_price)
    > over(partition by product_category_id order by cast (product_price as float)
    > rows between unbounded preceding and unbounded following) as c6
    > from products limit 10;
```

代码运行结果如图 4.26 所示。

product	category	price	c6
212	10	34.99	1999.99
201	10	79.99	1999.99
194	10	99	1999.99
195	10	99.95	1999.99
204	10	159.99	1999.99
207	10	159.99	1999.99
215	10	169.99	1999.99
216	10	189	1999.99
210	10	199.98	1999.99
211	10	199.99	1999.99

图4.26　行窗口统计示例代码运行结果

注意

UNBOUNDED PRECEDING 到 UNBOUNDED FOLLOWING 表示分组内所有行。

2. **范围窗口**

与行窗口相比，范围窗口不直接指定分组内第几行，而是取分组内“值在指定范围区间内的”行，该范围区间是通过用当前行的值加减指定的数字来决定的。目前，范围窗口只支持一个 ORDER BY 列。

【语法】

```
RANGE BETWEEN start_expr AND end_expr
```

其中，start_expr 与 end_expr 可设置为下列值。

- *n* PRECEDING/FOLLOWING：由当前行的值加（FOLLOWING）减（PRECEDING）*n*。
- CURRENT ROW：表示当前行的值。

例如，当前行的值为 3000，范围窗口定义如下：

```
PRECEDING 500 PRECEDING AND 1000 FOLLOWING
```

则范围区间为[2500,4000]。

示例 4-24

统计商品表中每个商品与其他商品价格差为 1 的商品总数。

关键代码：

```
hive> select product_id,product_price, count(product_price)
    > over(order by cast(product_price as float) range between 1 preceding and 1 following)
    > from products limit 20;
```

代码运行部分结果如图 4.27 所示。

```
product_id      product_price   _wcol0
1284    0       7
517     0       7
414     0       7
934     0       7
547     0       7
388     0       7
38      0       7
624     4.99    3
815     4.99    3
336     5       3
476     8       1
913     9.59    12
```

图4.27　范围窗口统计示例代码运行部分结果

在图 4.27 中，价格为“4.99”的商品对应的范围窗口为[3.99,5.99]，并且由图 4.27（按价格排序）可以直接观察到价格处于该范围区间的商品只有 3 个，编号依次为 624、815 与 336。

4.3.3 技能实训

使用窗口函数实现如下目标：

- 统计每日订单量；
- 完成每日销售额排行。

分析：

- 订单表中存在日期字段 order_date，由该字段进行分组；
- 销售额可以由订单明细表中 order_item_subtotal 求和得到；
- 对窗口函数的计算结果可以作为新的外层查询的子查询。

本章小结

- HQL SELECT 语句与 SQL 基本类似，包括 DISTINCT、FROM、WHERE、GROUP BY、SORT BY、ORDER BY、DISTRIBUTE BY、CLUSTER BY、LIMIT 等子句。
- HQL 支持 CTE，即可将查询作为临时表以便于共享。
- HQL 支持嵌套子查询，一般出现在 FROM 子句中，无嵌套级别限制。
- HQL 支持列匹配正则表达式和虚拟列。
- JOIN 分为内连接、外连接与交叉连接，其中外连接又分为左外连接、右外连接与全外连接。
- UNION ALL 用于合并多个具有相同结构的结果集。
- 排序包括全排序（ORDER BY）和局部排序（SORT BY、DISTRIBUTE BY、CLUSTER BY）。通常 DISTRIBUTE BY 和 SORT BY 会结合使用，如在同一列上的操作 CLUSTER BY=DISTRIBUTE BY+SORT BY。
- GROUP BY 用于分组，配合聚合函数使用。
- 窗口函数也可用于分组，但与 GROUP BY 不同。窗口函数是基于当前查询结果的分组操作，其比 GROUP BY 功能更强大、限制更小。
- 窗口函数包括排序、聚合及分析类函数。
- 聚合函数通常可支持窗口定义，可以进一步细分分组结果。

本章作业

一、简答题

1. HAVING 与 WHERE 有什么不同？
2. ORDER BY 与 SORT BY 的区别是什么？

二、编码题

1. 在零售数据中找出每个城市消费能力排名前两名顾客。
2. 统计每个商品大类的累积销售额。
3. 统计每个顾客每日的消费额。

第 5 章

Hive 函数与 Streaming

技能目标

- 掌握 Hive 常用内置函数。
- 理解自定义函数。
- 掌握用 Java 编写自定义函数。
- 掌握 Streaming 的使用。

本章任务

任务 1　应用内置函数。
任务 2　使用 Java 编写 Hive 自定义函数。
任务 3　使用 Streaming 实现数据处理。

本章资源下载

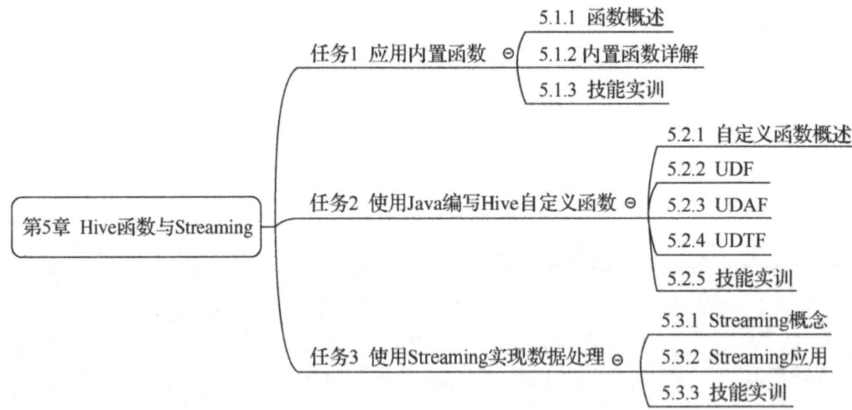

Hive 提供了很多函数供开发者使用，这些函数大多和关系型数据库中的函数类似，即通过计算输入然后输出计算结果。前面章节中已讲解过 Hive 的部分内置函数，可以说函数是 Hive 最常用的功能之一。除了使用内置函数，用户还可以通过自定义函数实现更加复杂的功能。本章将详细介绍 Hive 常用内置函数应用示例、3 种不同类型自定义函数以及不同于函数的另一种数据处理方式 Streaming。

任务 1 应用内置函数

【任务描述】

使用内置函数对零售数据进行操作。

- 转换顾客姓名为全大写。
- 对订单总金额进行四舍五入处理。
- 按月度统计订单数量。

【关键步骤】

（1）使用字符函数获取顾客完整姓名并进行大写转换。

（2）使用类型转换函数、聚合函数、数学函数实现金额四舍五入处理。

（3）使用日期函数和聚合函数实现订单的月度统计。

5.1.1 函数概述

从输入/输出的角度来看，函数可分为 3 类。

标准函数——以一行数据中的一列或多列数据作为输入参数且返回结果是一个值的函数。例如前面用过的 cast()函数，该函数可将输入数据的类型转换为新的数据类型。标准函数返回值只有一个，该值的数据类型既可以是基本类型，也可以是复杂类型。

聚合函数——以多行的零个到多个列的数据作为输入且返回单一值的函数。例如前面用过的 sum()、count()、max()等都属于聚合函数，其通常与 GROUP BY 子句结合使用。

表生成函数——接受零个或多个输入且产生多列或多行输出的函数。

1. **查看函数**

前面提到，SHOW FUNCTIONS 命令可以显示当前 Hive 会话中所加载的所有函数，包括内置函数与自定义函数。Hive 提供了许多内置函数，这些函数在 Hive 安装完成后即可使用，如图 5.1 所示。

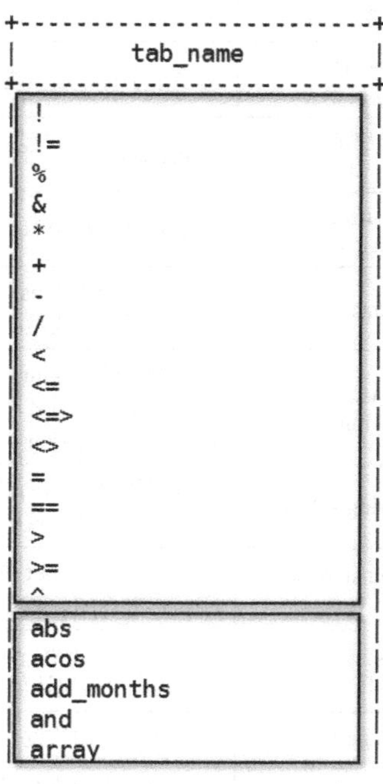

图5.1　Hive内置函数

图 5.1 中一部分为操作符，另一部分为函数。函数通常都有其自身的使用文档，可使用 DESCRIBE FUNCTION 命令来显示函数的简单描述，其中 DESCRIBE 可简写为 DESC。

【语法】

```
DESCRIBE FUNCTION <function_name>;
```

函数更多详细文档可通过增加 EXTENDED 关键字查看。

```
DESCRIBE FUNCTION EXTENDED <function_name>;
```

2. **调用函数**

通过在查询中调用函数名，并传入必要参数即可调用函数。函数调用可以在 SELECT 与 WHERE 子句中实现，具体包括下面 3 种典型情况。

- SELECT concat(col1,col2) AS x FROM table;
- SELECT concat('abc','def');
- SELECT * FROM table WHERE length(col)<10;

5.1.2　内置函数详解

Hive 提供了大量的内置函数供开发者使用，函数功能涵盖了字符处理、类型转换、聚合、数学计算等。

1. 字符函数

字符函数侧重于对字符串进行处理，常用字符函数如表 5-1 所示。

表 5-1 常用字符函数

返回值	函数签名	描述
string	concat(string|binary a, string|binary b,...)	对多个字符串或二进制字符码按参数顺序进行拼接
string	concat_ws(string SEP,string a,string b,...)	与 concat()类似，但第一个参数为指定的连接符
string	concat_ws(string SEP,array<string>)	与 concat_ws()类似，但拼接的是指定 array 中的元素
array<string>	split(string str,string pat)	按照正则表达式 pat 来分割字符串 str，并将分割后的结果以字符数组的形式返回
map<string,string>	str_to_map(text [,delimiter1,delimiter2])	将字符串转换为 map，第一个参数是需要转换的字符串，第二参数是键值对之间的分隔符，默认为逗号，第三个参数是键值之间的分隔符，默认为“=”
int	instr(string str,string substr)	查找字符串 str 中子字符串 substr 出现的位置，位置从 1 开始，如果查找失败则返回 0
string	substr(string a,int start[,int len])	对字符串 a 从 start 位置（索引为 1）开始截取长度为 len 的字符串并返回，其中 len 可选，未指定时表示截至末尾
int	locate(string substr,string str [,int pos])	查找字符串 str 中 pos 位置后字符串 substr 第一次出现的位置；pos 表示起始位置，未指定时为 0
int	length(string a)	返回字符串的长度
string	lower(string a)	将字符串中的所有字母转换成小写字母
string	upper(string a)	将字符串中的所有字母转换成大写字母

示例 5-1

在零售数据顾客表中转换顾客姓名为全大写。

关键代码：

```
hive> select upper(concat_ws(' ',customer_fname,customer_lname)) as fullname
    > from customers limit 10;
```

代码运行结果如图 5.2 所示。

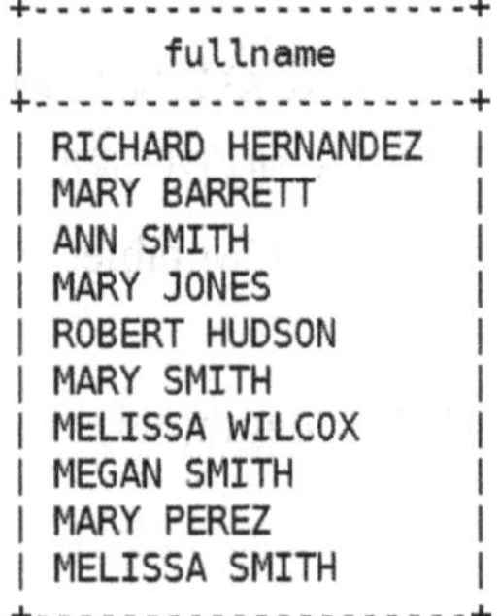

```
+--------------------+
|      fullname      |
+--------------------+
| RICHARD HERNANDEZ  |
| MARY BARRETT       |
| ANN SMITH          |
| MARY JONES         |
| ROBERT HUDSON      |
| MARY SMITH         |
| MELISSA WILCOX     |
| MEGAN SMITH        |
| MARY PEREZ         |
| MELISSA SMITH      |
+--------------------+
```

图5.2 字符函数示例代码运行结果

2. 类型转换函数

类型转换函数介绍如表 5-2 所示。

表 5-2　类型转换函数

返回值	函数签名	描述
binary	binary(string|binary)	将输入的值转换成二进制数
<type>	cast(expr as <type>)	将 expr 的类型转换成 type 类型，若转换失败则返回 NULL

3. 聚合函数

聚合函数包括前面已经学习过的 count、sum、max、min 及 avg 函数。与标准函数的不同之处在于，聚合函数是在一组多行数据中进行计算并返回单一值的函数。

4. 数学函数

数学函数通常针对数字类型进行操作。常用的数学函数介绍如表 5-3 所示。

表 5-3　常用数学函数

返回值	函数签名	描述
double	round(double a)	返回对 a 四舍五入的值，小数位为 0
double	round(double a,int d)	返回对 a 四舍五入并保留 d 位小数位的值
bigint	floor(double a)	向下取整，如 floor(3.14)=3
double	rand(),rand(int seed)	返回随机数，可指定随机因子 seed

示例 5-2

对订单总金额进行四舍五入，要求精度为小数点后两位。

关键代码：

```
hive> select order_id,round(sum(cast(order_items.order_ite_subtotal as float)),2)
    > from orders join order_items on orders.order_id=order_items.order_item_order_id
    > group by order_id limit 10;
```

代码运行结果如图 5.3 所示。

```
+-----------+---------+
| order_id  |   _c1   |
+-----------+---------+
| 1         | 299.98  |
| 10        | 651.92  |
| 100       | 549.94  |
| 1000      | 279.93  |
| 10000     | 209.97  |
| 10001     | 779.89  |
| 10002     | 129.99  |
| 10003     | 99.96   |
| 10004     | 659.86  |
| 10005     | 399.99  |
+-----------+---------+
```

图5.3　数学函数示例代码运行结果

5. 日期函数

常用日期函数介绍如表 5-4 所示。

表 5-4　常用日期函数

返回值	函数签名	描述
string	unixtime(bigint unixtime[,string format])	将时间的秒值转换成 format 格式，format 可为 "yyyy-MM-dd hh:mm:ss"、"yyyy-MM-dd hh:mm"、"yyyy-MM-dd" 等
bigint	unix_timestamp()	获取当前时刻本地时区的时间戳
bigint	unix_timestamp(string date)	将 "yyyy-MM-dd hh:mm:ss" 格式的时间字符串转为时间戳
string	to_date(string timestamp)	返回时间字符串的日期部分，如： to_date("1970-01-01 00:00:00")="1970-01-01"
int	year(string date)	返回时间字符串的年份部分，如： year("1970-01-01 00:00:00")=1970 year("1970-01-01")=1970
int	month(string date)	返回时间字符串的月份部分
int	day(string date)	返回时间字符串的天
int	hour(string date)	返回时间字符串的小时
int	minute(string date)	返回时间字符串的分钟
int	second(string date)	返回时间字符串的秒
int	datediff(string enddate,string startdate)	计算开始时间到结束时间的天数
string	date_add(string start date,int days)	从开始时间加上 days
string	date_sub(string start date,int days)	从开始时间减去 days

示例 5-3

统计月度订单数量。

关键代码：

```
hive> select from_unixtime(unix_timestamp(order_date),"yyyy-MM") as year_month,
    > count(order_id) from orders
    > group by from_unixtime(unix_timestamp(order_date),"yyyy-MM")
```

代码运行结果如图 5.4 所示。

```
+-------------+-------+
| year_month  |  _c1  |
+-------------+-------+
| 2013-07     | 1533  |
| 2013-08     | 5680  |
| 2013-09     | 5841  |
| 2013-10     | 5335  |
| 2013-11     | 6381  |
| 2013-12     | 5892  |
| 2014-01     | 5908  |
| 2014-02     | 5635  |
| 2014-03     | 5778  |
| 2014-04     | 5657  |
| 2014-05     | 5467  |
| 2014-06     | 5308  |
| 2014-07     | 4468  |
+-------------+-------+
```

图5.4　日期函数示例代码运行结果

6. 条件函数

Hive 提供的条件函数的介绍如表 5-5 所示。

表 5-5　条件函数

返回值	函数签名	描述
T	if(boolean testCondition,T valueTrue, T valueFalseOrNull)	如果 testCondition 为 true，则返回 valueTrue，否则返回 valueFalseOrNull
boolean	isnull(a)	如果 a 为 NULL，则返回 true，否则返回 false
boolean	isnotnull(a)	如果 a 不为 NULL，则返回 true，否则返回 false
T	nvl(T value,T defaultValue)	如果 value 为 NULL，则返回 defaultValue，否则返回 value
T	coalesce(T v1,T v2,...)	返回第一个非 NULL 的值，如： coalesce(NULL,1,2)=1
T	case a when b then c [when d then e]* [else f] end	当 a=b 时返回 c；当 a=d 时返回 e；否则返回 f
T	case when a then b [when c then d]* [else e] end	当 a=true 时返回 b；当 c=true 时返回 d；否则返回 e
void	asser_true(boolean condition)	用于测试，当 condition=false 时抛出异常

示例 5-4

根据商品价格将商品分为 3 个级别：0～100、100～200 及 200 以上，并分别统计各档商品个数。

关键代码：

```
hive> select level,count(*) from ( select * ,
    > case when product_price<100 then 1
    > when product_price between 100 and 200 then 2
    > else 3 end as level
    > from products ) as a group by level;
```

代码运行结果如图 5.5 所示。

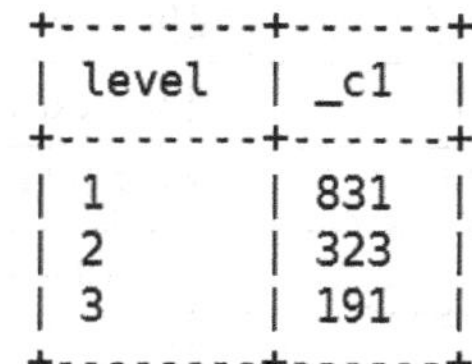

```
+--------+------+
| level  | _c1  |
+--------+------+
| 1      | 831  |
| 2      | 323  |
| 3      | 191  |
+--------+------+
```

图5.5　条件函数示例代码运行结果

7. 集合函数

集合函数用于对集合进行操作，其具体介绍如表 5-6 所示。

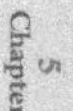

表 5-6　集合函数

返回值	函数签名	描述
int	size(map<K,V>)	返回 map 中键值对的个数
int	size(array<T>)	返回数组长度
array<K>	map_keys(map<K,V>)	返回 map 中的所有 key
array<V>	map_values(map<K,V>)	返回 map 中的所有 value
boolean	array_contains(array<T>,value)	查询 array 中是否包含 value
array<T>	sort_array(array<T>)	对数组进行排序

示例 5-5

针对前面章节中提到的雇员表 empdb.emp，使用集合函数查询其中所有员工所在部门的名称及总数。

关键代码：

```
hive> use empdb;
    > select name, map_keys(depart_title),size(depart_title) from emp;
```

代码运行结果如图 5.6 所示。

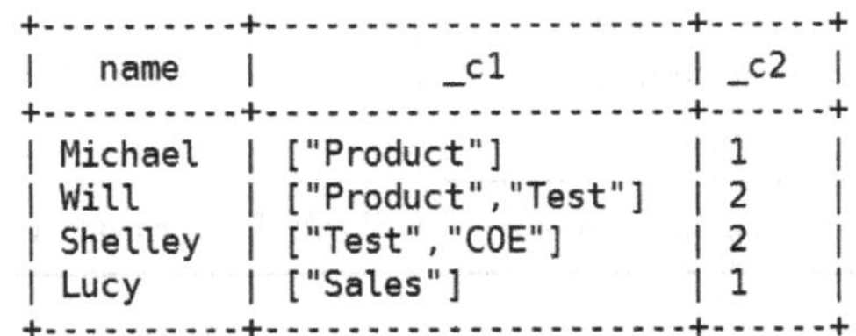

```
+----------+---------------------+------+
|   name   |         _c1         | _c2  |
+----------+---------------------+------+
| Michael  | ["Product"]         | 1    |
| Will     | ["Product","Test"]  | 2    |
| Shelley  | ["Test","COE"]      | 2    |
| Lucy     | ["Sales"]           | 1    |
+----------+---------------------+------+
```

图5.6　集合函数示例代码运行结果

8. 表生成函数

表生成函数可以将一个或多个输入值转换为多行输出，该输出可以作为表来使用。表生成函数具体介绍如表 5-7 所示。

表 5-7　表生成函数

返回值	函数签名	描述
N rows	explode(array<T>)	将 array 中的每个元素转成为表的一行，且行内须包含该元素
N rows	explode(map<K,V>)	将 map 中的每个键值对转成为表的一行，其中一个字段是键，另一个字段为值
N rows	posexplode(array<T>)	类似于 explode()，但额外返回一列包含了元素所在位置
N rows	stack(int n,v_1,...v_k)	将 k 列转换为 n 行，每行有 k/n 个字段，n 必须是常数
tuple	json_tuple(jsonStr,k1,k2,...)	从 JSON 字符串中获取多个键并将它们作为一个元组返回
tuple	parse_url_tuple(url,p1,p2,...)	从 URL 中抽取指定的参数内容，包括：HOST、PATH、QUERY、REF、PROTOCOL、AUTHORITY、FILE、USERINFO、QUERY:<KEY>
	inline(array<struct>)	将 array 中的结构体元素转换成行的形式

示例 5-6

使用表生成函数将如下数据生成为表。

- ['Apple', 'Orange', 'Mongo']
- {'A':'Apple', 'O':'Orange'}

关键代码：

```
hive> select explode(array('Apple', 'Orange', 'Mongo'));
    > select explode(map('A', 'Apple','O',' Orange '));
```

代码运行结果如图 5.7 所示。

```
0: jdbc:hive2://localhost:10000> select explode(array('Apple','Orange','Mongo'));
+---------+--+
|   col   |
+---------+--+
| Apple   |
| Orange  |
| Mongo   |
+---------+--+
3 rows selected (0.455 seconds)
0: jdbc:hive2://localhost:10000> select explode(map('A','Apple','O','Orange'));
+------+---------+--+
| key  |  value  |
+------+---------+--+
| A    | Apple   |
| O    | Orange  |
+------+---------+--+
2 rows selected (0.054 seconds)
```

图5.7　explode函数示例代码运行结果

示例 5-7

使用 stack 函数将字符值'a'、'b'、'c'、'd'分别生成为具有如下特征的表。

- 1 行 4 列
- 2 行 2 列
- 4 行 1 列

关键代码：

```
hive> select stack(1,'a','b','c','d');
    > select stack(2,'a','b','c','d');
    > select stack(4,'a','b','c','d');
```

代码运行结果如图 5.8 所示。

示例 5-8

将 json 字符串 "{"name":"Jason","age":"18"}" 转换为表。

关键代码：

```
hive> select json_tuple('{"name":"Jason","age":"18"}','name','age');
```

代码运行结果如图 5.9 所示。

```
0: jdbc:hive2://localhost:10000> select stack(1,'a','b','c','d');
+-------+-------+-------+-------+--+
| col0  | col1  | col2  | col3  |
+-------+-------+-------+-------+--+
| a     | b     | c     | d     |
+-------+-------+-------+-------+--+
1 row selected (0.051 seconds)
0: jdbc:hive2://localhost:10000> select stack(2,'a','b','c','d');
+-------+-------+--+
| col0  | col1  |
+-------+-------+--+
| a     | b     |
| c     | d     |
+-------+-------+--+
2 rows selected (0.446 seconds)
0: jdbc:hive2://localhost:10000> select stack(4,'a','b','c','d');
+-------+--+
| col0  |
+-------+--+
| a     |
| b     |
| c     |
| d     |
+-------+--+
4 rows selected (0.041 seconds)
0: jdbc:hive2://localhost:10000>
```

图5.8　stack函数示例代码运行结果

```
0: jdbc:hive2://localhost:10000> select json_tuple('{"name":"json","age":"18"}','name','age');
+-------+-----+--+
|  c0   | c1  |
+-------+-----+--+
| json  | 18  |
+-------+-----+--+
```

图5.9　json_tuple函数示例代码运行结果

更多 Hive 内置函数介绍请扫描二维码获取。

Hive 内置函数大全

5.1.3　技能实训

在商品表 products 中，使用字符函数对商品名称 product_name 进行截取并输出，要求商品名称不多于 x 个字符，其中 x 取第 10 个字符向后到首次出现的空格之间的字符个数。

分析：

- substr()可以实现字符截取；
- locate()可以实现 x 值的获取。

任务 2　使用 Java 编写 Hive 自定义函数

【任务描述】

在订单明细表中根据商品购买情况，给顾客一定的折扣。一个商品的购买数量与优惠间的关系如下。

- 购买 1 件，无折扣。
- 购买 2 件，9 折。
- 购买 3 件以上，8 折。

然后求出打折后的订单总额。

【关键步骤】

（1）实现自定义标准函数 UDF。

（2）实现自定义聚合函数 UDAF。

（3）实现自定义表生成函数 UDTF。

5.2.1　自定义函数概述

当 Hive 提供的内置函数无法满足应用需求时，就需要考虑使用自定义函数。按照类型的不同，Hive 提供的自定义函数可分为 3 类：自定义标准函数（User-Defined Function，UDF）、自定义聚合函数（User-Defined Aggregation Function，UDAF）以及自定义表生成函数（User-Defined Table-Generating Function，UDTF）。

Hive 中大量功能本质上都是通过自定义函数实现的。自定义函数通常使用 Java 语言编写，具体的编写方法稍后讨论。每一个自定义函数的开发过程包括如下步骤。

（1）编写自定义函数类。

（2）将自定义函数类编译为 JAR 文件。

（3）在 Hive 中注册 JAR 文件。

（4）利用 JAR 文件中的自定义函数类创建自定义函数。

（5）在 Hive 中调用自定义函数。

Hive 为不同类型的自定义函数提供了相应的 API（由 hive-exec 组件提供），如表 5-8 所示，在编写时应该首先确定是何种类型的自定义函数。

表 5-8　Hive 为不同类型的自定义函数提供的 API

函数类型	API
UDF	org.apache.hadoop.hive.ql.exec.UDF（已过时）
	org.apache.hadoop.hive.ql.udf.generic.GenericUDF
UDAF	org.apache.hadoop.hive.ql.exec.UDAF（已过时）
	org.apache.hadoop.hive.ql.udf.generic.AbstractGenericUDAFResolver
	org.apache.hadoop.hive.ql.udf.generic.GenericUDAFEvaluator
UDTF	org.apache.hadoop.hive.ql.udf.generic.GenericUDTF

项目工程 POM 文件依赖如下：

```
<dependencies>
        <dependency>
            <groupId>org.apache.hive </groupId>
            <artifactId>hive-exec</artifactId>
```

```
            <version>1.1.0-cdh5.14.2</version>
        </dependency>
        <dependency>
            <groupId>org.apache.hadoop</groupId>
            <artifactId>hadoop-common</artifactId>
            <version>2.6.0-cdh5.14.2</version>
        </dependency>
</dependencies>
```

5.2.2 UDF

自定义标准函数 UDF 有两种实现方式，分别为继承 UDF 类与 GenericUDF 类。如果自定义函数的输入与输出均为基本数据类型（Writable 类型），则继承 UDF 类即可；针对复杂数据类型（如 ARRAY、MAP、STRUCT），则须继承 GenericUDF 类。

1. 简单方式

继承 org.apache.hadoop.hive.ql.exec.UDF 较为简单，重写 evaluate()方法即可。该方法支持不同参数个数的重载。

示例 5-9

使用 UDF 将字符串变为全小写。

实现步骤：

（1）编写自定义函数类 MyLower 并继承 UDF 类；

（2）将 MyLower 类编译为 JAR 文件；

（3）注册 JAR 文件；

（4）创建函数；

（5）调用函数。

关键代码：

```
public class MyLower extends UDF {
    public Text evaluate(final Text s) {
        if (s == null) { return null; }
        return new Text(s.toString().toLowerCase());
    }
    public Text evaluate(final Text s1,final Text s2) {
        if (s1 == null || s2 == null) { return null; }
        return new Text((s1.toString()+s2.toString()).toLowerCase());
    }
    //测试 UDF
    public static void main(String[] args) {
        MyLower udf=new MyLower();
        System.out.println(udf.evaluate(new Text("Hello")));
        System.out.println(udf.evaluate(new Text("Hello"),new Text("World")));
    }
}
```

进入项目工程目录，执行如下命令。

```
$ mvn package
$ hive
hive> add jar target/hive-udf-api-1.0-SNAPSHOT.jar;
    > create temporary funtion mylower as 'hive.ch05.udf.MyLower';
    > show functions 'my.*';
    > select mylower('HELLO');
    > select mylower('HELLO','World');
```

使用“create temporary funtion”创建的是临时函数，当函数测试成功后，可使用“create funtion”创建永久函数。使用简单方式创建 UDF 的代码运行结果如图 5.10 所示。

```
0: jdbc:hive2://localhost:10000> show functions 'my.*';
+-----------+--+
| tab_name  |
+-----------+--+
| mylower   |
+-----------+--+
1 row selected (0.023 seconds)
0: jdbc:hive2://localhost:10000> select mylower('HELLO');
+--------+--+
|  _c0   |
+--------+--+
| hello  |
+--------+--+
1 row selected (0.056 seconds)
0: jdbc:hive2://localhost:10000> select mylower('HELLO','World');
+-------------+--+
|     _c0     |
+-------------+--+
| helloworld  |
+-------------+--+
1 row selected (0.041 seconds)
```

图5.10 使用简单方式创建UDF

2. 复杂方式

org.apache.hadoop.hive.ql.udf.generic.GenericUDF 要求开发者手动管理函数的输入参数的类型和数量。继承该类要求实现以下 3 个。

（1）abstract ObjectInspector initialize(ObjectInspector[] arguments)

该函数在 evaluate()函数前被调用，用于指定函数的输入参数与返回值的解析器。函数中 ObjectInspector 数组用于描述函数的参数，其返回也是 ObjectInspector 类型，负责解析 evaluate()函数返回的对象。ObjectInspector 为底层对象类型提供了一个统一的接口，以便 Hive 中不同的对象能以相同的方式加以实现。

ObjectInspector 需要手动管理，其子抽象类和接口包括：PrimitiveObjectInspector、ListObjectInspector、MapObjectInspector、StructObjectInspector、UnionObjectInspector，其中 PrimitiveObjectInspector 用于解析基本数据类型。在本质上，对 ObjectInspector 进行手动管理就是为实际数据的类型选择了相应的 ObjectInspector。

（2）abstract Object evaluate(GenericUDF.DeferredObject[] arguments)

该函数类似 UDF 类的 evaluate()函数，用于接收实际参数并返回结果。arguments 是

DeferredObject 类型，而 ObjectInspector 通过 DeferredObject 可获取实际输入参数对象，也就是说 evaluate()函数接收到的数据由 ObjectInspector 负责解析。同时，evaluate()函数返回的对象也会被 ObjectInspector 解析。

（3）abstract String getDisplayString(String[] children)

该函数用于展示调试信息。

示例 5-10

使用 GenericUDF 实现商品折扣计算。商品打折说明：购买 1 件无折扣，购买 2 件打 9 折，购买 3 件以上打 8 折。

关键代码：

```
public class Discount extends GenericUDF {
    IntObjectInspector num;
    @Override
    public ObjectInspector initialize(ObjectInspector[] arguments)
            throws UDFArgumentException {
        // 指定输入参数解析器，当前函数只传一个参数
        this.num=(IntObjectInspector)arguments[0];
        // 指定返回值解析器，当前函数返回 double 类型
        return PrimitiveObjectInspectorFactory.javaDoubleObjectInspector;
    }
    @Override
    public Object evaluate(DeferredObject[] arguments)
            throws HiveException {
        // 获得商品购买数量
        // 首先通过 DeferredObject.get()获取实际参数，然后由 ObjectInspector 对其进行解析
        int i=this.num.get(arguments[0].get());
        double discount=1.0;
        // 计算折扣
        switch (i){
            case 1:discount=1.0;break;
            case 2:discount=0.9;break;
            case 3:discount=0.8;break;
            default:discount=0.8;break;
        }
        return discount;
    }
    @Override
    public String getDisplayString(String[] children) {
        return "discount(num)";
    }
}
hive> user retail_db;
    > add jar target/hive-udf-api-1.0-SNAPSHOT.jar;
    > create temporary discount as 'hive.ch05.udf.Discount';
```

```
> select order_item_order_id,order_item_id,order_item_quantity,
> discount(cast(order_item_quantity as int)) as discount
> from order_items limit 10;
```

代码运行结果如图 5.11 所示。

order_item_order_id	order_item_id	order_item_quantity	discount
1	1	1	1.0
2	2	1	1.0
2	3	5	0.8
2	4	1	1.0
4	5	2	0.9
4	6	5	0.8
4	7	3	0.8
4	8	4	0.8
5	9	1	1.0
5	10	5	0.8

图5.11　使用复杂方式创建UDF示例代码运行结果

5.2.3 UDAF

自定义聚合函数也有 2 种实现方式，即简单方式与复杂方式。

1. **简单方式**

自定义聚合函数可通过继承 org.apache.hadoop.hive.ql.exec.UDAF 类，并在内部包含多个实现了 UDAFEvaluator 的静态类加以实现。UDAFEvaluator 是封装了 UDAF 计算逻辑的接口，每个 UDAF 都需要计算器（Evaluator），其数目视不同需求而定，例如 max()和 min()聚合函数具有用于整数、字符串和其他类型的计算器，而 avg()只有用于 double 类型的计算器。

由于该方式存在性能缺陷，故其已被弃用，接下来着重介绍另一种方式：复杂方式。

2. **复杂方式**

实现 UDAF 的复杂方式涉及 AbstractGenericUDAFResolver 和 GenericUDAFEvaluator 两个类。

- org.apache.hadoop.hive.ql.udf.generic.AbstractGenericUDAFResolver
- org.apache.hadoop.hive.ql.udf.generic.GenericUDAFEvaluator

主要实现步骤如下。

（1）继承 AbstractGenericUDAFResolver，重写 getEvaluator()方法。

（2）继承 GenericUDAFEvaluator 抽象类，为 AbstractGenericUDAFResolver 的 getEvaluator()方法生成 Evaluator 实例，进而实现如下方法。

- init()：由 Hive 调用，用于初始化一个 UDAF Evaluator 类的实例。其包括指定聚合的模式（Mode）和各个阶段输入参数与返回值的数据解析器。
- getNewAggregationBuffer()：返回一个新的 AggregationBuffer 对象以存储临时聚合结果。

➢ iterate()：在 Map 阶段处理一行新的数据并将处理结果存储到 AggregationBuffer 对象中。

➢ terminatePartial()：返回当前聚合的内容。

➢ merge()：将 terminatePartial()返回的中间部分聚合结果合并到当前聚合中。

➢ terminate()：在 Reducer 阶段输出最终结果。

➢ reset()：重置 AggregationBuffer 对象以使 AggregationBuffer 对象可重用。

聚合的 Mode 决定了在 MapReduce 期间对输入值进行计算时 Evaluator 中哪些方法将被调用。GenericUDAFEvaluator.Mode 是静态枚举类，其包括如下枚举值。

➢ Mode.PARTIAL1：表示在 map 阶段将会调用 iterate()和 terminatePartial()。

➢ Mode.PARTIAL2：表示在 map 端的 combine 阶段将会调用 merge()和 terminatePartial()。

➢ Mode.FINAL：表示在 reduce 阶段将会调用 merge()和 terminate()。

➢ Mode.COMPLETE：表示 MapReduce 只有 map 而没有 reduce，将会调用 iterate()和 terminate()。

完整的 UDAF 逻辑即是一次 MapReduce 过程，如果有 Mapper 与 Reducer，则会经历 PARTITAL1 和 FINAL。如果还有 Combiner 则将经历 PARTITAL1、PARTITAL2 和 FINAL。如果只有 Mapper 那么就只经历 COMPLETE 阶段。

示例 5-11

自定义聚合函数以实现 sum()函数功能，并统计商品打折后订单总金额。

关键代码：

```
public class MySum extends AbstractGenericUDAFResolver {
    @Override
    public GenericUDAFEvaluator getEvaluator(TypeInfo[] parameters)
            throws SemanticException {
        return new GenericUDAFSumDouble();
    }
    public static class GenericUDAFSumDouble extends GenericUDAFEvaluator{
        // 输入参数解析器
        private PrimitiveObjectInspector inputOI;
        @Override
        public ObjectInspector init(Mode m, ObjectInspector[] parameters) throws HiveException {
            super.init(m, parameters);
            inputOI=(PrimitiveObjectInspector)parameters[0];
            // 指定返回值解析器，当前函数返回 double 类型
            return PrimitiveObjectInspectorFactory.writableDoubleObjectInspector;
        }
        // 存储聚合结果
        static class SumDoubleAgg implements AggregationBuffer {
            double sum;
        }
        @Override
        public AggregationBuffer getNewAggregationBuffer() throws HiveException {
```

```
            SumDoubleAgg agg=new SumDoubleAgg();
            reset(agg);
            return agg;
        }
        @Override
        public void reset(AggregationBuffer agg) throws HiveException {
            ((SumDoubleAgg)agg).sum=0;
        }
        @Override
        public void iterate(AggregationBuffer agg, Object[] parameters) throws HiveException {
            merge(agg, parameters[0]);
        }
        @Override
        public Object terminatePartial(AggregationBuffer agg) throws HiveException {
            return terminate(agg);
        }
        @Override
        public void merge(AggregationBuffer agg, Object partial) throws HiveException {
            SumDoubleAgg myagg = (SumDoubleAgg) agg;
            myagg.sum += PrimitiveObjectInspectorUtils.getDouble(partial, inputOI);
        }
        @Override
        public Object terminate(AggregationBuffer agg) throws HiveException {
            SumDoubleAgg result=(SumDoubleAgg)agg;
            return new DoubleWritable(result.sum);
        }
    }
}
hive> add jar target/hive-udf-api-1.0-SNAPSHOT.jar;
   > create temporary function mysum as 'hive.ch05.udf.MySum';
   > select order_item_order_id,mysum(order_item_subtotal)
   > from order_items group by order_item_order_id;
```

代码运行部分结果如图 5.12 所示。

order_item_order_id	subtotal	discount_price
1	299.98	299.98
10	651.9200000000001	641.924
100	549.94	477.94800000000004
1000	279.93	269.93399999999997
10000	209.97000000000003	201.972
10001	779.89	661.908
10002	129.99	129.99
10003	99.96	89.964
10004	659.86	587.884
10005	399.99	359.99

图5.12　使用复杂方式创建UDAF示例代码运行部分结果

5.2.4 UDTF

由于 ObjectInspector 的作用，在 UDF 和 UDAF 中返回值可以是任意类型的对象，但只能返回一行一列数据，对返回多行多列的情况还须使用表成生函数。

在 Hive 中使用内置表生成函数 explode()很容易将数组[1,2,3,4,5]输出为表格结构，但有些情况下生成表的数据是未知的，explode()将无法满足要求，此时即须通过自定义 UDTF 来满足上述要求。

编写 UDTF 需要继承 org.apache.hadoop.hive.ql.udf.generic.GenericUDTF 类，并实现以下 3 个方法。

（1）StructObjectInspector initialize(ObjectInspector[] args)

UDTF 首先调用该方法进行生成表结构初始化。该方法中 args 为输入参数解析器，返回的 StructObjectInspector 定义了生成表的行信息，包括所有字段名与字段类型。

（2）void process(Object[] args)

初始化完成后 UDTF 将会调用该方法，其中 args 为函数输入的实际参数。在该方法中每调用一次 forward(Object obj)方法就会产生一行表数据。forward()方法为 GenericUDTF 已实现的方法，无须重写。如果表数据为多个列，应该使用 Object[]保存多个列的值，然后将值传递给 forward()。

（3）close()

当所有的行都生成后，调用该方法进行代码清理。

示例 5-12

编写函数 range(x,y)，当指定整数 x 与 y 时，能够返回由 x 到 y 之间的所有数字形成的表。

关键代码：

```
public class Range extends GenericUDTF {
    @Override
    public StructObjectInspector initialize(ObjectInspector[] argOIs)
            throws UDFArgumentException {
        // 定义生成表的行信息
        ArrayList<String> fieldNames = new ArrayList<String>();
        ArrayList<ObjectInspector> fieldOIs = new ArrayList<ObjectInspector>();
        fieldNames.add("col1");
        fieldOIs.add(PrimitiveObjectInspectorFactory.javaIntObjectInspector);
        // 返回生成表的行信息
        return ObjectInspectorFactory.getStandardStructObjectInspector(fieldNames,fieldOIs);
    }
    @Override
    public void process(Object[] args) throws HiveException {
        int start=Integer.parseInt(args[0].toString());
        int end=Integer.parseInt(args[1].toString());
        for(int i=start;i<=end;i++)
```

```
        {
            // 只有一列
            Object[] forwardObj=new Object[1];
            forwardObj[0]=i;
            this.forward(forwardObj);
        }
    }
    @Override
    public void close() throws HiveException {}//执行清理
}
```

```
hive> add jar target/hive-udf-api-1.0-SNAPSHOT.jar;
> create temporary function range as 'hive.ch05.udf.Range';
> select range(1,5);
```

代码执行结果如图 5.13 所示。

```
0: jdbc:hive2://localhost:10000> add jar /root/workspace/hive/hive-udf-api-1.0-SNAPSHOT.jar;
No rows affected (0.066 seconds)
0: jdbc:hive2://localhost:10000>  create temporary function range as 'hive.ch05.udf.Range';
No rows affected (0.051 seconds)
0: jdbc:hive2://localhost:10000>  select range(1,5);
+-------+--+
| col1  |
+-------+--+
| 1     |
| 2     |
| 3     |
| 4     |
| 5     |
+-------+--+
5 rows selected (0.124 seconds)
```

图5.13 创建UDTF示例代码运行结果

5.2.5 技能实训

编写一个 UDF 以识别字符串中的所有不同单词，即输入如下内容：

“Hello Hive,Hello Hadoop,Hello World!”

经过 UDF 函数处理后返回：“Hello,Hive,Hadoop,World”。

分析：

- 选择一种实现 UDF 的方式；
- 确定输入类型；
- 确定输出类型；
- 编写 Java 代码以实现不同单词的识别。

任务 3 使用 Streaming 实现数据处理

【任务描述】

结合 Linux Shell 常用命令与 Streaming 完成数据操作。

【关键步骤】

（1）掌握 Streaming 的概念。

（2）了解常用 Streaming 应用操作。

（3）了解更多 Streaming 应用操作。

5.3.1 Streaming 概念

除了 Java 自定义函数外，还可以使用其他语言（如 Shell、Python、Perl 等）扩展 Hive 的数据处理功能。Streaming 为这些语言提供了一种处理数据的方式，为外部进程开启了一个 I/O 管道。数据通过 I/O 管道被传给外部进程，该进程通过标准输入（管道）读取数据，然后通过标准输出（管道）写回结果数据。

受管道中的数据序列化和反序列效率的影响，Streaming 的执行效率比 UDF 的效率要低，并且很难调试整个程序，但是对于不熟悉 Java 的 Hive 用户来说，Streaming 是非常方便的。

Streaming 可以实现恒等变换、类型改变、投影变换、操作转换、分布式内存使用、一行转多列、聚合计算、排序（CLUSTERBY、DISTRIBUTE BY、SORT BY）等。

Hive 提供多个语法来使用 Streaming，具体包括 MAP()、REDUCE()、TRANSFORM()等。

【语法】

```
SELECT TRANSFORM(cols...) USING 'script' [AS (cols...)] FROM table;
```

通常情况下使用 Streaming 的语法是 TRANSFORM，MAP 和 REDUCE 可以看作是 TRANSFORM 的别名，仅作为提高可读性存在。需要注意的是，使用 MAP 和 REDUCE 语法时，Hive 并不保证一定在 Map 或 Reduce 阶段中调用脚本。

语法中“script”既可以是命令也可以是脚本文件。如果其是脚本文件，则须先使用“add file”命令将其加入工程。

使用 AS 指定输出的列名，对应至脚本输出的列，列分隔符为“\t”。如果省略列名，则固定输出两列，第一列为 key，其余列均为 value。

5.3.2 Streaming 应用

1. 恒等变换

恒等变换，又称不动变换，是指数据经过处理而未发生变化。例如 Shell 中“/bin/cat”命令可以将传递给它的数据直接输出。

示例 5-13

使用 cat 命令查看顾客表 customer_id 与 customer_fname。

关键代码：

```
hive> select transform(customer_id,customer_fname) using '/bin/cat' as id,name from customers limit 10;
```

代码运行结果如图 5.14 所示。

```
+-----+----------+
| id  |   name   |
+-----+----------+
| 1   | Richard  |
| 2   | Mary     |
| 3   | Ann      |
| 4   | Mary     |
| 5   | Robert   |
| 6   | Mary     |
| 7   | Melissa  |
| 8   | Megan    |
| 9   | Mary     |
| 10  | Melissa  |
+-----+----------+
```

图5.14　恒等变换示例代码运行结果

2. 类型改变

TRANSFORM 返回字段类型默认为字符串类型，使用 AS 子句可以将其类型转换为其他数据类型。

示例 5-14

使用 cat 命令查看顾客表 customer_id 与 customer_fname，并将 customer_id 表中的数据类型转换为 double。

关键代码：

```
hive> select transform(customer_id,customer_fname)
    > using '/bin/cat' as (id int,name string)
    > from customers limit 10;
```

代码运行结果如图 5.15 所示。

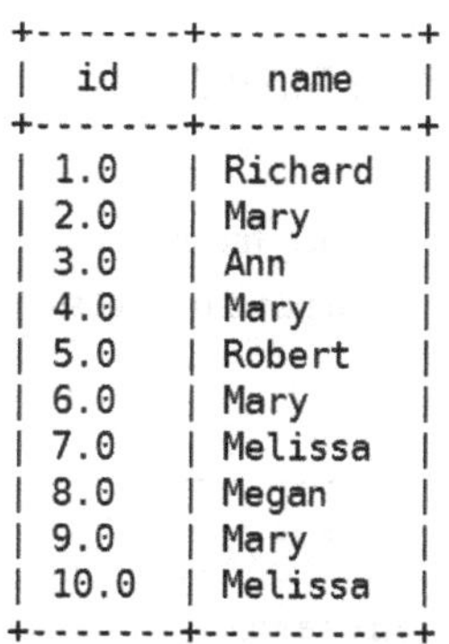

```
+-------+----------+
|  id   |   name   |
+-------+----------+
| 1.0   | Richard  |
| 2.0   | Mary     |
| 3.0   | Ann      |
| 4.0   | Mary     |
| 5.0   | Robert   |
| 6.0   | Mary     |
| 7.0   | Melissa  |
| 8.0   | Megan    |
| 9.0   | Mary     |
| 10.0  | Melissa  |
+-------+----------+
```

图5.15　类型改变示例代码运行结果

注意

当进行数据类型转换时，所有字段都应被指定类型。

3. 投影变换

Shell cut 命令是选取命令，即在分析一段数据之后从中选出所需部分。一般来说，选取信息是针对“行”进行的。

示例 5-15

取出顾客表的第一列。

关键代码：

```
hive> select transform(*) using '/bin/cut -f1' as id,others from customers limit 10;
```

其中“cut –f1”表示取第一个字段。上述代码运行结果如图 5.16 所示。

```
+-----+---------+
| id  | others  |
+-----+---------+
| 1   | NULL    |
| 2   | NULL    |
| 3   | NULL    |
| 4   | NULL    |
| 5   | NULL    |
| 6   | NULL    |
| 7   | NULL    |
| 8   | NULL    |
| 9   | NULL    |
| 10  | NULL    |
+-----+---------+
```

图5.16 投影变换示例代码运行结果

4. 操作转换

Shell sed 命令是一个面向数据流的非交互式编辑器，可以首先向其中输入数据流，然后按照用户指定的方式进行编辑，最后将编辑结果输出。

例如，使用 sed 命令替换字符串方法为：“sed s/待替换的字符串/新字符串/文件名”。

示例 5-16

使用 sed 将顾客 Mary 的名字改为 MARY。

关键代码：

```
hive> select transform(customer_id,customer_fname)
    > using '/bin/sed s/Mary/MARY/' as id,name from customers limit 10;
```

代码运行结果如图 5.17 所示。

```
+-----+----------+
| id  |   name   |
+-----+----------+
| 1   | Richard  |
| 2   | MARY     |
| 3   | Ann      |
| 4   | MARY     |
| 5   | Robert   |
| 6   | MARY     |
| 7   | Melissa  |
| 8   | Megan    |
| 9   | MARY     |
| 10  | Melissa  |
+-----+----------+
```

图5.17 操作转换示例代码运行结果

Streaming 应用

更多 Streaming 应用请扫描二维码查看。

5.3.3　技能实训

使用 Streaming 将顾客表中的顾客名称转换为全大写。

分析：

➢ 运用 Shell sed 完成操作转换；

➢ 运用 sed '/s[a-z]/\u&/g'可将匹配的小写字符转换为大写字符。

本章小结

➢ Hive 函数分为 3 类：标准函数、聚合函数和表生成函数。

➢ Hive 提供了大量的内置函数，可实现字符操作、类型转换、数学运算、日期处理、集合操作、聚合统计、表生成等操作。

➢ Hive 提供了 3 种函数自定义方式：UDF、UDAF 和 UDTF，它们均须继承相应的 API 并实现其中的方法。

➢ Streaming 为 Hive 提供了其他非 Java 语言以扩展数据处理的能力，典型的语言包括 Shell 与 Python。

本章作业

一、简答题

1．Hive 的 3 类函数的区别是什么？

2．简述 UDF、UDAF 及 UDTF 的开发过程。

二、编码题

1．编写一个 UDF，实现字符串类型转 MAP 类型。输入字符串格式为："k1=v1,k2=v2,…"。

2．编写一个 UDAF，实现 array_sum(ARRAY<int>)函数，即使其能够对分组内某列数据（ARRAY<int>类型）进行求和统计。

3．编写一个 UDTF，将"1、2、3、4、5、6、7、8、9"输出为三行三列的表，列名不限。

第 6 章

Hive 视图与索引

技能目标

➢ 理解 Hive 视图的概念。
➢ 掌握 Hive 视图的常用操作。
➢ 了解 Hive 索引的概念。
➢ 掌握 Hive 索引的常用操作。

本章任务

任务 1　创建并管理零售商店的顾客表和订单表视图。
任务 2　建立零售商店顾客表索引。

本章资源下载

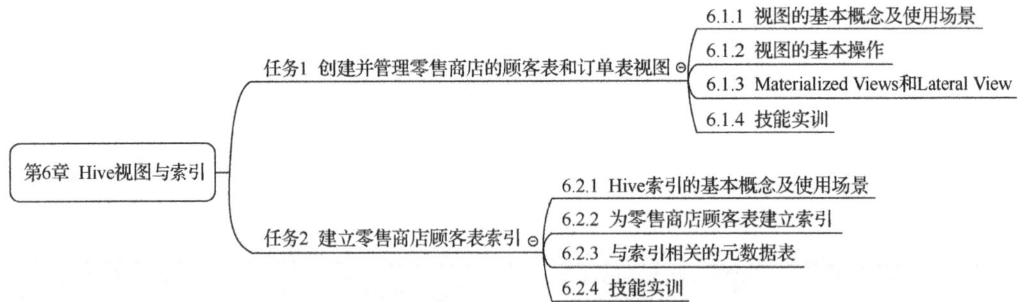

视图和索引在 MySQL 等关系型数据库中是非常重要的功能，视图可以简化用户的操作，索引可以在一定程度上提高数据查询的效率。在 Hive 数据仓库中，同样也提供了视图和索引功能。本章主要针对视图和索引的概念及用法进行讲解。

任务 1 创建并管理零售商店的顾客表和订单表视图

【任务描述】

建立基于零售商店顾客表和订单表的视图，并管理之。

【关键步骤】

（1）了解视图的基本概念及使用场景。

（2）使用 Hive 视图命令管理视图。

6.1.1 视图的基本概念及使用场景

1. **视图的基本概念**

视图是隐藏了子查询、连接查询等操作的（简化后的）逻辑结构，它由从数据库的真实表中选取出来的数据组成，是一个与真实表不同的虚拟表。视图只保存定义而不存储数据，行和列的数据均来自视图所引用的表，它们将在使用视图时动态生成。如果引用表的列被删除，则会造成视图的错误。

从 Hive 数据仓库内部来看，视图是由使用 SELECT 语句从一张和多张表中获取的数据组成的虚拟表。从 Hive 数据仓库外部来看，视图类似于一张表，表的查询操作都可以应用于视图。当一个查询引用一个视图时，视图所定义的查询语句和用户定义的查询语句会组合在一起，然后供 Hive 执行查询操作。从逻辑上讲，Hive 会先执行视图，然后使用执行结果执行后续查询。注意，在 Hive 中视图是只读的，操作人员不能向视图中插入或是装载数据。

2. **视图的使用场景及优势**

（1）视图的使用场景

对于一些真实表而言，它们不希望未授权的用户查看其具有特殊安全性的行或列，此时可以通过建立视图，选取能提供给用户的列，并授权用户查看。比如，在 retail_db 数据库的 customers 表中，“customer_password”是顾客的隐私信息，其不应该提供给用户访问，这时就可以使用视图来限制数据访问权限以达到保护顾客信息不被随意查询的目的，代码如下。

```
CREATE VIEW customers_ view AS SELECT customer_fname,customer_lname,
customer_email,customer_state,customer_city
FROM customers;
```

上述代码中“CREATE VIEW”是创建视图的命令，后续会对其做详细讲解。上述代码通过创建视图，将 retail_db.customers 表中的“customer_password”字段隐藏，即只提供给用户查询视图 customers_ view 的权限，进而保证了顾客信息中的敏感字段不会被随意查询。

针对查询语句非常复杂的场景，如多表关联查询及多重子查询，可以使用视图来代替复杂的查询语句。在 retail_db 数据库中，如下的查询就是一个具有嵌套子查询的查询：

```
FROM（
        SELECT c.customer_id,c.customer_email,c.customer_state,c.customer_city,
        cancat(c.customer_lastname," ",c.customer_fname) AS customer_name,
        o.order_date,o.order_status
        FROM customers   c
        JOIN orders o ON (c.customer_id = o.order_customer_id)
        WHERE c.customer_fname = 'Mary'
）a SELECT a. customer_name WHERE a.order_status='COMPLETE';
```

Hive 查询语句中，多表连接查询及嵌套查询非常常见，可以面向上述嵌套了子查询的查询建立视图，代码如下：

```
CRAETE VIEW customer_orders_view AS
SELECT c.customer_id,c.customer_email,c.customer_state,c.customer_city,
cancat(c.customer_lastname," ",c.customer_fname) AS customer_name,
o.order_date,o.order_status
FROM customers   c
JOIN orders o ON (c.customer_id = o.order_customer_id)
WHERE c.customer_fname = 'Mary';
```

上述代码面向嵌套了子查询的查询语句创建了视图 customer_orders_view，这样就可以像操作表一样来操作 customer_orders_view。之前的嵌套子查询语句可以简写如下。

```
SELECT customer_name FROM customer_orders_view
WHERE order_status='COMPLET';
```

由此可见，使用视图极大程度简化了之前的查询语句。

（2）视图的优势

使用视图可以简化用户对数据的理解，用户可以将注意力集中在其关心的数据上而不是全部的数据。针对不能供用户访问的一些敏感数据，比如顾客信息表中的“customer_password”，可以使用视图过滤掉该字段。

使用视图可以降低查询复杂度。当查询语句中存在复杂的连接查询及子查询时，可以使用视图将该查询语句进行分隔，进而降低查询复杂度。

6.1.2 视图的基本操作

Hive 视图的操作命令包括 CREATE、SHOW、DROP、ALTER 等，主要用于进行 Hive 视图创建、查看、修改（功能有局限）和删除等操作。

1. 创建视图

【语法】

CREATE VIEW [IF NOT EXISTS] [db_name.]view_name [(column_name [COMMENT column_comment], …)]

[COMMENT view_comment]

[TBLPROPERTIES (property_name = property_value, …)]

AS SELECT …;

“CREATE VIEW”是创建视图命令，其属性“COMMENT”和“TBL PROPERTIES”与创建表时的含义相同，“AS SELECT”用于指定查询语句，查询语句可以是复杂的嵌套子查询也可以是关联查询，查询结果将作为视图中的内容。

示例 6-1

创建基于零售商店顾客信息表和订单表的 customer_orders_view 视图。

需要创建的 customer_orders_view 视图应包含的字段说明如表 6-1 所示。

表 6-1　customer_orders_view 视图字段

数据表表字段	说明	来源表
customer_fname	顾客名字	customer
customer_lname	顾客姓氏	customer
order_id	订单 ID	orders
order_date	订单日期	orders
order_status	订单状态	orders

关键代码：

```
create view customer_orders_view(
customer_fname,
customer_lname,
order_id,
order_date,
order_status)
comment 'this is customer_orders_view'
as select c.customer_fname,c.customer_lname,o.order_id,o.order_date,o.order_status
from retail_db.customers c
left join retail_db.orders o on c.customer_id = o.order_customer_id;
```

注意

创建的视图与其他表或视图不能同名，用户可使用关键字“IF NOT EXISTS”避免犯该错误。

本示例中，视图的列名是在其创建的时候指定的，可以为每一列添加“COMMENT”属性，单独为视图的每个列添加描述信息。Hive 支持创建视图时不提供列名称。如果没有提供列名称，视图的列将自动引用 SELECT 语句中的列名称。

Hive 视图可以包含“ORDER BY”和“LIMIT”子句，值得注意的是，如果某个查询引用了视图，同时也包含了“ORDER”或“LIMIT”子句，那么视图的子句的执行优先级将要高于查询的子句，例如一个视图“v”指定了“LIMIT 5”，则查询语句“SELECT * FROM v LIMIT 10”最多返回 5 行记录。

2. **查看视图**

在 Hive 2.2.0 版本之前，查看已经创建的视图只能使用 SHOW TABLES 命令，而从 Hive 2.2.0 版本开始，Hive 提供了 SHOW VIEWS 命令，可以列出当前数据库中所有的视图。本书使用的 Hive 版本是 hive-1.1.0-cdh5.14.2，故将采用 SHOW TABLES 命令来查看视图。

【语法】

```
SHOW TABLES [IN/FROM database_name] [LIKE 'pattern_with_wildcards'];
```

其中“IN/FROM”关键字用于指定数据库，“LIKE”关键字用于进行模糊匹配。

示例 6-2

查看数据库中的视图。

关键代码：

```
show tables;
```

代码运行结果如图 6.1 所示。

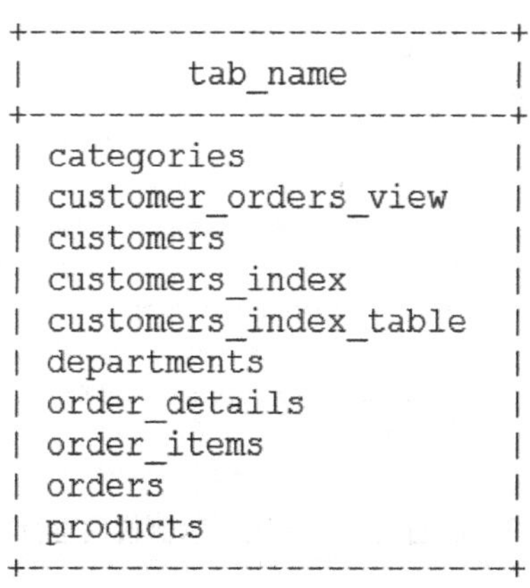

图6.1 查看数据库中的视图示例代码运行结果

从图 6.1 中可以看出，创建的视图 customer_orders_view 已经在列表中了，但是如何辨别哪些是表、哪些是视图呢？从 Hive 2.2.0 开始，使用“SHOW VIEWS”和“SHOW TABLES”命令可以分别列出视图和表；在 Hive 2.2.0 之前，可以使用“DESC FORMATTED”命令显示视图的元数据信息，输出信息中的“Detailed Table Information”将存在一个“Table Type”字段中，字段的值为“VIRTUAL_VIEW”。

示例 6-3

查看 customer_orders_view 视图的详细信息。

关键代码：

```
desc formatted customer_orders_view;
```

代码运行结果如图 6.2 所示。

```
| # Detailed Table Information                          | NULL
                               |
| Database:                                             | retail_db
                               |
| Owner:                                                | anonymous
                               |
| CreateTime:                                           | Wed Dec 26 11:45:35 CST 2018
                               |
| LastAccessTime:                                       | UNKNOWN
                               |
| Protect Mode:                                         | None
                               |
| Retention:                                            | 0
                               |
| Table Type:                                           | VIRTUAL_VIEW
                               |
| Table Parameters:                                     | NULL
```

图6.2　视图详细信息查看示例代码运行结果

Hive 提供了查看视图定义的方式，语法如下。

【语法】

```
SHOW CREATE TABLE view_name;
```

示例 6-4

查看示例 6-1 创建的视图 customer_orders_view 的定义。

关键代码：

```
show create table customer_orders_view;
```

代码运行结果如图 6.3 所示。

```
OK
CREATE VIEW `retail_db.customer_orders_view` AS sele
ct `c`.`customer_fname`,`c`.`customer_lname`,`o`.`or
der_date`,`o`.`order_status` from `retail_db`.`custo
mers` `c` left join `retail_db`.`orders` `o` on `c`.
`customer_id` = `o`.`order_customer_id` where `c`.`c
ustomer_lname` = 'Smith'
Time taken: 0.047 seconds, Fetched: 1 row(s)
```

图6.3　视图定义信息查看示例代码运行结果

从图 6.3 中可以看出，视图的定义信息与创建视图的语句是一致的。

视图创建完成后，可以像表一样使用查询语句查询视图。

示例 6-5

查看视图 customer_orders_view 中的前 10 条数据。

关键代码：

```
select * from customer_orders_view limit 10;
```

代码运行结果如图 6.4 所示。

```
Richard Hernandez         22945   2013-12-13 00:00:00      COMPLETE
Mary     Barrett 15192    2013-10-29 00:00:00    PENDING_PAYMENT
Mary     Barrett 33865    2014-02-18 00:00:00    COMPLETE
Mary     Barrett 57963    2013-08-02 00:00:00    ON_HOLD
Mary     Barrett 67863    2013-11-30 00:00:00    COMPLETE
Ann      Smith   22646    2013-12-11 00:00:00    COMPLETE
Ann      Smith   23662    2013-12-19 00:00:00    COMPLETE
Ann      Smith   35158    2014-02-26 00:00:00    COMPLETE
Ann      Smith   46399    2014-05-09 00:00:00    PROCESSING
Ann      Smith   56178    2014-07-15 00:00:00    PENDING
```

图6.4 查看视图前10条数据示例代码运行结果

3. 修改视图

使用 ALTER 命令可以修改 Hive 视图的元数据信息，包括视图的属性和定义。

（1）修改视图属性

【语法】

```
ALTER VIEW [db_name].view_name SET TBLPROPERTIES table_properties;
```

其中“table_properties”的格式如下。

```
(property_name=property_value,property_name=property_value,….)
```

table_properties 可以包含多组键值对，各键值对用逗号分隔。

类似于 ALTER TABLE 语句，可以使用 ALTER VIEW 语句将用户的元数据添加到视图中。

示例 6-6

修改视图 customer_orders_view 的属性，为视图添加有效期限（term）和用户组（user_group）属性。

关键代码：

```
alter view customer_orders_view set tblproperties ('term' = 'three month','user_group'='IT department');
```

上述代码执行完成后，可以使用命令“desc formatted customer_orders_view;”查看视图的详细信息。

代码运行结果如图 6.5 所示。

```
Table Parameters:
        comment                 this is customer_orders_view
        last_modified_by        anonymous
        last_modified_time      1545803810
        term                    three month
        transient_lastDdlTime   1545803810
        user_group              IT department
```

图6.5 查看为视图添加的属性信息示例代码运行结果

从视图的属性信息中可以看出，“Table Parameters”字段属性里面已经记录了本示例为视图添加的两个属性（term 和 user_group）。

（2）修改视图定义

Hive 在 0.11 版本后，提供了“ALTER VIEW AS SELECT”命令来修改视图的定义。

【语法】

```
ALTER VIEW [db.name].view_name AS select_statement;
```

该语法与创建视图的语法类似，不同之处在于修改视图要求视图必须存在。

示例 6-7

修改视图 customer_orders_view 的定义，要求其只获取顾客姓氏为“Smith”的数据。

关键代码：

```
alter view customer_orders_view
as select c.customer_fname,c.customer_lname,o.order_date,o.order_status
from retail_db.customers c
left join retail_db.orders on c.customer_id = o.order_customer_id
where c.customer_lname='Smith';
```

修改完成后，可以使用命令“show create table customer_orders_view;”查看视图的定义信息。代码运行结果（即新的视图定义信息）如图 6.6 所示。

```
CREATE VIEW `customer_orders_view` AS select `c`.`customer_fname`,`c`.`cus
tomer_lname`,`o`.`order_date`,`o`.`order_status` from `retail_db`.`custome
rs` `c` left join `retail_db`.`orders` `o` on `c`.`customer_id` = `o`.`ord
er_customer_id` where `c`.`customer_lname` = 'Smith'
Time taken: 0.06 seconds, Fetched: 1 row(s)
```

图6.6　新的视图定义信息

从新的视图定义信息中可以看出，视图创建语句已经修改成了只获取姓氏为“Smith”的数据。

4. 删除视图

DROP VIEW 会删除指定视图的元数据，当删除的视图被其他视图依赖时，Hive 不会给出警告，但是依赖被删除视图的其他视图将会变成无效视图，必须由用户删除或重新创建。

【语法】

```
DROP VIEW [IF NOT EXISTS] [db_name.]view_name;
```

示例 6-8

删除 customer_orders_view 视图。

关键代码：

```
drop view customer_orders_view;
```

视图删除完成后，可以使用“show tables”命令来验证视图“customer_orders_view”已经不存在。

注意

Hive 0.7.0 之后，如果视图不存在，DROP 命令会报错，可以使用两种方式去解决该问题。

① 在删除视图命令中添加“IF NOT EXISTS”关键字。

② 修改 Hive 配置变量“hive.exec.drop.ignorenonexistent”的值为“true”。

6.1.3 Materialized Views 和 Lateral View

1. **物化视图**（Materialized Views）

（1）Materialized Views 概述

Materialized Views 是对数据的一个快照，在执行复杂的表连接或聚集等耗时较多的查询操作时，可以使用 Materialized Views 预先计算并保存查询结果，如此一来，在真正执行查询时就可以快速查询到结果。Materialized Views 本身属于数据库的高级功能，在传统的关系型数据库中基本上都已经实现了 Materialized Views。Materialized Views 与普通视图的区别在于：Materialized Views 会存储数据，具有和表一样的特征；而普通视图不存储数据，是只有表结构的虚拟表。

Hive 3.0 以后，增加了对 Materialized Views 的支持，Hive 的 Materialized Views 是基于 Apache Calcite 项目实现的。

（2）Materialized Views 操作

目前，Materialized View 支持的操作还比较局限，主要包括 CREATE、SHOW、DESCRIBE、DROP 等。由于本书使用的 Hive 版本是 hive-1.1.0-cdh5.14.2，不支持 Materalized View 的操作，所以本节只介绍 Materalized View 的语法。Materialized Views 操作案例请扫描二维码获取。

Materalized View 操作案例

① 创建 Materialized View

【语法】

```
CREATE MATERIALIZED VIEW [IF NOT EXISTS] [db_name.]materialized_view_name
  [DISABLE REWRITE]
  [COMMENT materialized_view_comment]
  [PARTITIONED ON (col_name, ...)]
  [
    [ROW FORMAT row_format]
    [STORED AS file_format]
      | STORED BY 'storage.handler.class.name' [WITH SERDEPROPERTIES (...)]
  ]
  [LOCATION hdfs_path]
  [TBLPROPERTIES (property_name=property_value, ...)]
AS
<query>;
```

② 查看所有的 Materialized View

【语法】

```
SHOW MATERIALIZED VIEWS [IN database_name] ['identifier_with_wildcards'];
```

③ 查看指定 Materialized View 的信息

【语法】

```
DESCRIBE [EXTENDED | FORMATTED] [db_name.]materialized_view_name;
```

④ 删除 Materialized View

【语法】

```
DROP MATERIALIZED VIEW [db_name.]materialized_view_name;
```

2. **侧视图**（Lateral View）

（1）Lateral View 概述

Lateral View 一般与 UDTF 结合使用，UDTF 会为每个输入行生成零个或者多个输出行。Lateral View 会将 UDTF 应用于基表的每一行，然后将结果输出行对应地连接到输入行。在前面的章节中已经介绍过 UDTF，在使用 UDTF 时，Hive 只允许对拆分字段进行访问，比如以 empdb 数据库中 emp 表为例，执行下面的语句是正确的。

```
SELECT explode(work_place) FROM empdb.emp;
```

但是如果使用下列语句：

```
SELECT name,work_place explode(work_place) FROM empdb.emp;
```

则会提示错误信息：

"UDTF's are not supported outside the SELECT clause,nor nested in expressions"。

SELECT 只能选择 UDTF 表达式中的列，但是在实际场景中，经常需要拆分完某个字段后将其与别的字段一起输出，比如在 emp 表中需要对"work_place"进行拆分，然后将拆分完的字段与对应的"name"字段一起输出，这就需要使用 Lateral View 来实现。Lateral View 可以将拆分的单个字段数据与原始表关联上，可以理解为 Lateral View 是原始表的 INNER JOIN，关联键就是原始表的行号。Lateral View 最典型的使用案例是与 explode()函数的结合使用。

（2）Lateral View 操作

① 单个 Lateral View

【语法】

```
LATERAL VIEW udtf(expression) tableAlias AS columnAlias (',' columnAlias)*
```

其中，"expression"可以是 ARRAY、MAP 等类型的字段名，也可以是其他函数，如 split()函数；"tableAlias"是视图别名，必须被指定；"columnAlias"是侧视图在新生成的数据集上的列别名。

示例 6-9

在雇员信息表中用 Lateral View 和 explode()函数按不同的工作地统计员工个数。

分析：

在雇员信息表中，工作地存储在字段 work_place 下，数据类型为 array<string>。explode()函数可以进行"行转列"操作，然后 Lateral View 可以将转换后的结果与原表对应行进行连接，进而达到统计效果。

关键代码：

```
select name,work_place,wps from empdb.emp lateral view explode(work_place)
work_place_single as wps;
```

代码运行结果如图 6.7 所示。

```
+-----------+------------------------+------------+
|   name    |       work_place       |    wps     |
+-----------+------------------------+------------+
| Michael   | ["Montreal","Toronto"] | Montreal   |
| Michael   | ["Montreal","Toronto"] | Toronto    |
| Will      | ["Montreal"]           | Montreal   |
| Shelley   | ["New York"]           | New York   |
| Lucy      | ["Vancouver"]          | Vancouver  |
+-----------+------------------------+------------+
```

图6.7　Lateral View操作示例代码运行结果

从图 6.7 中可以看出，第三列的数据已经变成了单个的工作地，我们可以很方便按工作地来统计员工的个数，代码如下。

```
select wps,count(*) from (
select name,work_place,wps from empdb.emp lateral view explode(work_place)
work_place_single as wps
)
A group by wps;
```

注意

上述代码中的“work_place_single”是 UDTF 生成表的名称，必不可少。

② Mutilple Lateral Views

在示例 6-9 中，FROM 子句只使用了一个 LATERAL VIEW 子句，Hive 支持 FROM 子句中有多个 LATERAL VIEW 子句。LATERAL VIEW 可以引用其左边表中的任何列。LATERAL VIEW 子句将按照它们出现的顺序被应用，使用方式如下。

```
SELECT * FROM exampleTable
LATERAL VIEW explode(col1) myTable1 AS myCol1
LATERAL VIEW explode(myCol1) myTable2 AS myCol2;
```

示例 6-10

在示例 6-9 基础上，展示出雇员的技能和分数。

关键代码：

```
select name,wps,skill,score from empdb.emp
lateral view explode(work_place) work_place_single as wps
lateral view explode(skills_score) sks as skill,score;
```

代码运行结果如图 6.8 所示。

```
+----------+-----------+---------+--------+
|   name   |    wps    |  skill  | score  |
+----------+-----------+---------+--------+
| Michael  | Montreal  | DB      | 80     |
| Michael  | Toronto   | DB      | 80     |
| Will     | Montreal  | Perl    | 85     |
| Shelley  | New York  | Python  | 80     |
| Lucy     | Vancouver | Sales   | 89     |
| Lucy     | Vancouver | HR      | 94     |
+----------+-----------+---------+--------+
```

图6.8　Mutilple Lateral View操作示例代码运行结果

③ Outer Lateral Views

当 UDTF 不产生任何行时，如 explode()函数作用的列为空时不产生任何行，在这种情况下，元数据行是不会出现在结果中的。比如编写如下代码：

```
select name,loc from empdb.emp lateral view explode(split(null, ',')) a as loc;
```

代码运行结果如图 6.9 所示。

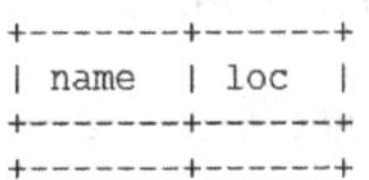

图6.9 Lateral View返回空结果

上述代码中的 explode 函数的参数会对“null”进行切分操作，操作不会产生行，基础表 empdb.emp 中有数据，但是最终输出的是空结果，这显然是不符合实际应用场景的。如果想要元数据行出现在结果中，必须使用“OUTER”关键字，类似于左外连接。在 Hive 0.12.0 版本后，用户可以指定可选的 OUTER 关键字来解决返回空结果的问题。UDTF 的列输出空结果时会用 NULL 值来代替。

示例 6-11

使用 OUTER 关键字解决 UDTF 返回空结果问题。

关键代码：

```
select name,work_place,loc from empdb.emp
lateral view outer explode(split(null, ',')) a as loc;
```

代码运行结果如图 6.10 所示。

```
+----------+-------------------------+-------+
|   name   |       work_place        |  loc  |
+----------+-------------------------+-------+
| Michael  | ["Montreal","Toronto"]  | NULL  |
| Will     | ["Montreal"]            | NULL  |
| Shelley  | ["New York"]            | NULL  |
| Lucy     | ["Vancouver"]           | NULL  |
+----------+-------------------------+-------+
```

图6.10 Outer Lateral View操作示例代码运行结果

6.1.4 技能实训

使用 Hive Shell 命令行来完成对零售商店数据库 retail_db 中表的视图管理。基于订单表、订单明细表和产品表的视图 orders_products_items_view 的字段说明如表 6-2 所示。

表 6-2 orders_products_items_view 视图字段

数据表表字段	字段说明	来源表
order_id	订单 ID	order_items
order_date	订单日期	orders
order_status	订单状态	orders
product_name	产品名称	products
product_price	产品价格	products
product_quantity	购买产品数量	order_items

关键步骤：

（1）创建视图 orders_products_items_view；

（2）查询视图 orders_products_items_view 是否存在；

（3）为视图 orders_products_items_view 添加属性；

（4）删除 orders_products_items_view 视图；

（5）查询视图 orders_products_items_view 是否存在。

任务 2 建立零售商店顾客表索引

【任务描述】

为零售商店顾客表建立索引并管理之。

【关键步骤】

（1）了解 Hive 索引的基本概念及使用场景。

（2）为零售商店顾客表建立索引。

6.2.1 Hive 索引的基本概念及使用场景

1. Hive 索引的基本概念

在关系型数据库中，使用索引是为了加速对数据表中某些列的查询。索引就是将数据库表中一列或者多列的值进行排序存储，并用索引表记录字段的索引和偏移量，以方便查询索引列时能快速定位到对应的数据。索引类似于图书的目录，我们可以根据目录中标记的页码对各章节进行快速定位。

Hive 从 0.7.0 版本开始支持索引功能，但是 Hive 索引与关系型数据库中的索引有所区别：Hive 索引不支持键（主键和外键）的操作。Hive 索引的设计目的同样是提高表中某些列的查询速度。如果没有索引，当执行查询语句“SELECT * FROM test_table WHERE key = 'xx'”时，Hive 会加载整个 test_table 表或者分区数据并处理所有行；如果“key”列存在索引，则只需要加载和处理文件的一部分，这在一定程度上可以减少 MapReduce 任务中读取数据块的数量。Hive 在指定的列上建立索引时会产生索引表（Hive 的一张物理表），该表包含的字段有：索引列的值、该值对应的 HDFS 文件路径、该值在文件中的偏移量。在对索引字段进行查询时，Hive 首先会额外产生一个 MapReduce Job，其将根据对索引列的过滤条件从索引表中过滤出索引列的值对应的 HDFS 文件路径及偏移量，并输出到 HDFS 临时文件中；然后根据文件中的 HDFS 路径和偏移量过滤原始文件，并生成新的逻辑块（split）以作为 MapReduce Job 的 split，进而达到不用扫描全表的目的。比如，查询语句“SELECT * FROM test_table WHERE key = 'xx'”不对 key 建立索引时，执行流程如图 6.11 所示。

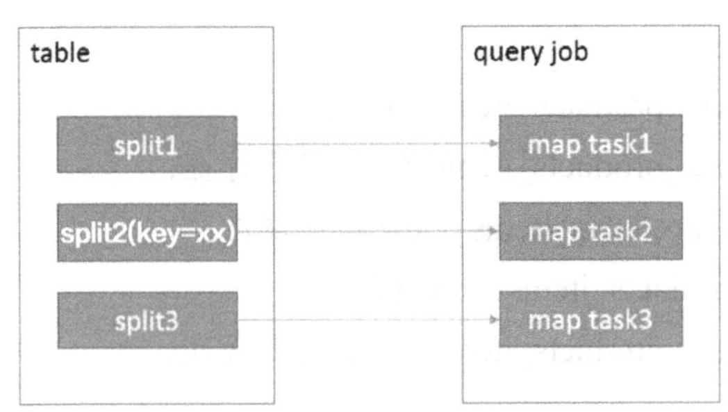

图6.11 不使用索引的Hive执行流程

从图 6.11 可以看出，表 customers 的每一个 split 都会用一个 map task 去扫描，但其实只有 split2 中存在想要的结果数据，因此 map task1 和 map task3 的执行属于资源浪费。

如果要求上面的查询语句使用索引，则须在 customers 表的“customer_fname”上建立索引，执行流程如图 6.12 所示。

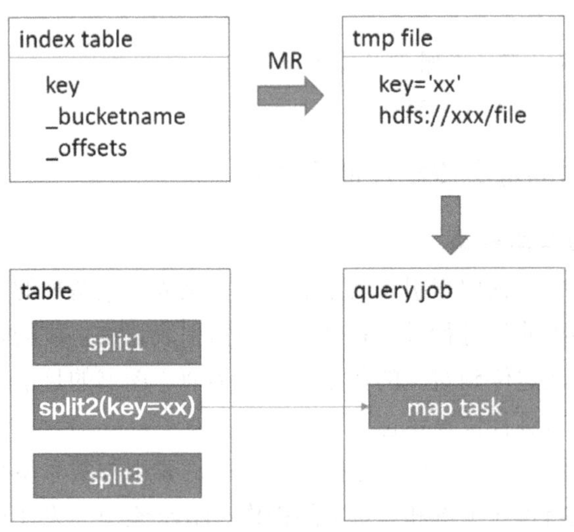

图6.12 使用索引的Hive执行流程

从图 6.12 中可以看出，查询提交后，Hive 会先用一个 MapReduce 扫描索引表，并从索引表中找出 key 值为'xx'的记录，以获取对应的 HDFS 文件名和偏移量；然后直接定位到文件中的偏移量，并用一个 map task 完成查询，如此即可减少查询的数据量。

2. Hive 索引的使用场景

（1）对于在查询中经常被当作 WHERE 子句的判断条件的列而言，其可以建立索引。比如，在用户信息表中查询数据时我们通常会将姓名作为判断条件，在姓名列上建立索引可以提高查询效率。

（2）值不经常更新的列或者值是几个枚举值的列可以建立索引。在这种类型的列上建立索引，可以避免数据更新后重建索引的操作，也可以减少建立索引所占用的空间。

值得注意的是，Hive 中的索引和关系型数据库中的一样，均须通过仔细评估后才能使用。维护索引需要额外的存储空间，同时创建索引页需要消耗计算资源。索引不是建立的越多越好，我们需要在建立索引权衡其为查询带来好处和因此需要付出的代价。

（3）从 Hive 3.0 版本开始，索引将被移除，但 Hive 又提供了与索引类似的功能。

➢ 使用带有自动重写功能的物化视图，Hive 3.0 增加了对物化视图的支持。

➢ 选择列式的文件格式（如 Parquet、ORC 等），它们可以进行选择性扫描，甚至可以跳过整个文件快。

需要说明的是，Hive 索引、分区、分桶都是 Hive 的优化手段。索引是使用额外的存储空间来换取查询时间，其不对数据库进行分隔。分区是将整个数据库按照分区字段拆分成多个小数据库，并使它们对应 HDFS 上的不同文件夹。分桶是按照列的哈希函数对数据进行分隔，并将分隔所得对应到 HDFS 上的不同文件。通常情况下，分桶应用于对分区中的数据进行进一步拆分成桶。

6.2.2 为零售商店顾客表建立索引

Hive 索引的操作命令包括 CREATE、SHOW、DROP、ALTER 等，主要用于进行 Hive 索引创建、查询、修改和删除等操作。

1. 创建索引

【语法】

```
CREATE INDEX index_name
ON TABLE base_table_name (col_name, …)
AS 'index.handler.class.name'
[WITH DEFERRED REBUILD]
[IDXPROPERTIES (property_name=property_value, …)]
[IN TABLE index_table_name]
[PARTITION BY (col_name,…)]
[
        [ ROW FORMAT …] STORED AS …
        | STORED BY …
]
[LOCATION hdfs_path]
[TBLPROPERTIES (…)]
[COMMENT "index comment"];
```

其中：

➢ “ROW FORMAT”“STORED AS”“STORED BY”“LOCATION”“PARTITION BY”等子句的使用方式可以参照创建表的语法。

➢ AS 子句指定了索引处理器，也就是一个索引接口的 Java 类。Hive 内置的索引处理器有 CompactIndexHandler 和 BitMap。

➢ WITH DEFERRED REBUILD 用于延迟重建标识，即如果用户指定了 DEFERRED REBUILD，新索引将呈现空白状态，用户可以在任何时间进行第一次索引创建或者使用 ALTER INDEX 对索引进行重建。

➢ IN TABLE index_table_name 用于指定索引表的名字，其为可选值。如果不指定，

Hive 会默认生成索引表名。

示例 6-12

为顾客表 customers 的“customer_fname”列建立索引。

分析：

在顾客表中，“customer_fname”字段被当作 WHERE 子句的查询判断条件，所以可以在该列上建立索引。

关键代码：

```
create index fname_index on table customers(customer_fname)
as 'org.apache.hadoop.hive.ql.index.compact.CompactIndexHandler'
with deferred rebuild
in table customer_index_table
comment 'customers index by fanme';
```

索引创建完成后，Hive 会在当前使用的数据库下创建索引表“customers_index_ table”。使用 show tables 命令可查看索引表如图 6.13 所示。

```
+------------------------+
|        tab_name        |
+------------------------+
| categories             |
| customer_orders_view   |
| customers              |
| customers_index        |
| customers_index_table  |
| departments            |
| order_details          |
| order_items            |
| orders                 |
| products               |
+------------------------+
```

图6.13 索引表

注意

使用 Hive 索引前须先设置两个参数：hive.optimize.index.filter（是否启用索引的自动使用）和 hive.optimize.index.groupby（是否使用聚合索引优化 group by 操作），它们的默认值都是 false。在使用 Hive 索引进行查询优化时，须将 hive.optimize.index.filter 设置为 true。更多 Hive 索引相关配置介绍请扫描二维码了解。

2. 重建索引

使用 AlTER INDEX 命令可以重建索引，该命令主要用于重建使用“WITH DEFERRED REBUILD”语句创建的索引，如果指定了分区，则仅重建指定分区的索引。

Hive 索引相关配置介绍

【语法】

```
ALTER INDEX index_name ON table_name [PARTITION partition_spec] REBUILD;
```

Hive 目前不支持索引原始表或其某个分区数据发生改变时自动触发重建索引这一机制。

示例 6-13

重建 customers 表上的 fname_index。

分析：

为顾客表创建索引时使用了“WITH DEFERRED REBUILD”关键字，这使得在当前数据库下建立的表 customers_index_table 是空的，因此需要重建索引并更新索引表。

关键代码：

```
alter index fname_index on customers rebuild;
```

重建索引以后，索引表 customers_index_table 中才会记录字段对应的 HDFS 文件路径及偏移量，索引表详细信息如图 6.14 所示。

```
customer_fname   _bucketname  _offsets
Alan     hdfs://cluster1/data/retail_db/customers/customers.csv  [28372,28559,86655,126299,140984,232689,...]
Albert  hdfs://cluster1/data/retail_db/customers/customers.csv  [610935,622679,772412,839013,973215,1015005,...]
Alexander   hdfs://cluster1/data/retail_db/customers/customers.csv  [8418,30705,33510,66940,77676,96260,...]
```

图6.14　索引表详细信息

3. 显示索引

索引创建后，可以使用 SHOW INDEX 命令来查看索引。

【语法】

SHOW [FORMATTED] (INDEX|INDEXES) ON table_with_index [(FROM|IN) db_name];

其中，关键字“FORMATTED”是可选的，使用该关键字可以格式化输出，即使输出的结果中包含有列名称。使用“INDEXES”关键字可以列出多条索引信息。

示例 6-14

查看顾客表上的索引信息。

关键代码：

```
show formatted index on customers;
```

代码运行结果如图 6.15 所示。

idx_name	tab_name	col_names	idx_tab_name	idx_type	comment
fname_index	customers	customer_fname	customer_index_table	compact	customes index by customer_fname

图6.15　顾客表上的索引信息

4. 删除索引

为原始表创建索引后，Hive 中会相应地创建一个索引表，该表是 Hive 的物理表，但是 Hive 不允许用户直接使用 DROP TABLE 语句来删除该索引表，而是必须使用 DROP INDEX 语句去删除索引。

【语法】

DROP INDEX [IF EXISTS] index_name ON table_name;

其中，IF EXISTS 是可选的，其可在索引不存在时抛出错误信息。

示例 6-15

删除 customers 表上建立的索引 fname_index。

关键代码：

```
drop index if exists fname_index on table customers;
```

注意

如果原始表被删除，那么其对应的索引和索引表也会被删除。同样地，如果原始表的某个分区被删除，该分区对应的分区索引也会被删除。

6.2.3 与索引相关的元数据表

在 Hive 元数据表中，与索引相关的数据表主要包括 TBLS、IDXS、INDEX_PARAMS 等，其中 TBLS 表在前面的章节中已经介绍过了，此处不再赘述。

1. **元数据表 IDXS**

IDXS 表中包含每个索引创建的实例信息，并且记录了与元数据表 TBLS 的关联信息。

元数据表 IDXS 的各字段说明如表 6-3 所示。

表 6-3 元数据表 IDXS 的各字段说明

元数据表字段	说明	示例数据
INDEX_ID	索引 ID	3
CREATE_TIME	创建时间	1545118376
DEFERRED_REBUILD	延迟重建标识	空
INDEX_HANDLER_CLASS	索引处理类	org.apache.hadoop.hive.ql.index.compact.CompactIndexHandler
INDEX_NAME	索引名字	order_item_product_id_index
INDEX_TBL_ID	索引表的 ID	788
LAST_ACCESS_TIME	最后访问时间	1545118376
ORIG_TBL_ID	原始表的 ID	782
SD_ID	序列化配置信息	791

2. **元数据表 INDEX_PARAMS**

INDEX_PARAMS 表中包含每个索引的属性信息，比如创建时间戳、最新修改用户、最新修改时间等。

元数据表 INDEX_PARAMS 的各字段说明如表 6-4 所示。

表 6-4 元数据表 INDEX_PARAMS 的各字段说明

元数据表字段	说明	示例数据
INDEX_ID	索引 ID	3
PARAM_KEY	属性名	base_timestamp
PARAM_VALUE	属性值	1545047366911

示例 6-16

查看 Hive 中与索引相关的元数据表。

注意

由于示例 6-15 中删除了创建的索引，所以此处需要重新创建索引，可参照示例 6-12。

关键代码：

进入 MySQL 服务命令行界面。

（1）查看 IDXS 表

```
mysql> select * from IDXS\G;
```

代码运行结果如图 6.16 所示。

```
*************************** 1. row ***************************
          INDEX_ID: 17
       CREATE_TIME: 1545960825
  DEFERRED_REBUILD: □
INDEX_HANDLER_CLASS: org.apache.hadoop.hive.ql.index.compact.CompactIndexHandler
        INDEX_NAME: fname_index
      INDEX_TBL_ID: 816
  LAST_ACCESS_TIME: 1545960825
       ORIG_TBL_ID: 775
             SD_ID: 828
1 row in set (0.00 sec)
```

图6.16　IDXS表信息

（2）查看元数据表 INDEX_PARAMS

mysql> select * from INDEX_PARAMS\G;

代码运行结果如图 6.17 所示。

```
*************************** 1. row ***************************
   INDEX_ID: 17
  PARAM_KEY: base_timestamp
PARAM_VALUE: 1545019114120
*************************** 2. row ***************************
   INDEX_ID: 17
  PARAM_KEY: comment
PARAM_VALUE: customes index by customer_fname
*************************** 3. row ***************************
   INDEX_ID: 17
  PARAM_KEY: last_modified_by
PARAM_VALUE: anonymous
*************************** 4. row ***************************
   INDEX_ID: 17
  PARAM_KEY: last_modified_time
PARAM_VALUE: 1545960900
*************************** 5. row ***************************
   INDEX_ID: 17
  PARAM_KEY: transient_lastDdlTime
PARAM_VALUE: 1545960900
5 rows in set (0.01 sec)
```

图6.17　INDEX_PARAMS表信息

6.2.4　技能实训

使用 Hive Shell 命令行方式完成对零售商店数据库 retail_db 中表的索引的管理。

➢　为 retail_db 数据库中的 orders 表的 order_date 列建立索引并查看之。

➢　使用 MySQL 在 Hive 元数据库中查询的索引相关的元数据表。

关键步骤：

（1）使用 CREATE INDEX 命令在 orders 表的 order_date 列上创建索引“order_date_index”；

（2）使用 ALTER INDEX 重建索引；

（3）使用 SHOW INDEX 命令查看 orders 表上建立的索引信息；

（4）使用 SQL 语句查询相关元数据的信息；

（5）使用 DROP INDEX 命令删除建立在 order 表上的索引；再次查看建立在 orders 表上的索引，验证其是否删除成功。

本章小结

➢ Hive 视图是一个逻辑概念。在实际工作中，我们可以为基于多张物理表的复杂查询建立视图，进而给用户提供简单的视图操作方式。

➢ 使用侧视图可以解决“Hive UDTF 只允许对拆分字段进行访问”这一问题。

➢ 物化视图会预先计算并保存表连接或聚集等操作的结果，可以提高数据查询效率。

➢ Hive 索引的目的是提高 Hive 表指定列的查询速度，减少 MapReduce 任务中需要读取的数据块的个数。

本章作业

一、简答题

1. 简述 Hive 视图的使用场景。

2. 简述 Hive 索引的使用场景。

二、编码题

1. 基于 retail_db 数据库中的数据表（如 customers、orders、order_items、products 等）建立视图，要求其能够显示顾客的名字、订单日期、订单状态、购买的商品及购买商品的数量。

2. 为用户信息表 user_info 建立索引。用户信息表包含两个字段，分别为 id 和 name。user_info 表字段说明如表 6-5 所示。

表 6-5　user_info 表字段说明

表字段名称	类型	说明
id	int	用户 id
name	string	用户姓名

需求：

➢ 为用户信息表的 id 列建立索引，并设置索引延迟重建；

➢ 完成索引的重建、查看（包括查看表上的索引和索引表的元数据信息）和删除操作。

3. 使用 Lateral View 和 explode()函数统计互联网上特定广告在网页上展示的次数。网页信息表 pageAds 有 pageid 和 adid_list 两个字段，其字段说明如表 6-6 所示。

表 6-6　pageAds 表字段说明

表字段名称	类型	说明	示例数据
pageid	string	网页名称	contact_page
adid_list	array<int>	该网页上显示的广告数	[3,4,5]

第 7 章

Hive 调优

技能目标

- 熟悉 Hive 性能调优。
- 掌握 Hive 性能调优策略。

本章任务

任务 1　熟悉 Hive 性能调优策略。
任务 2　解决 Hive 数据倾斜问题。
任务 3　Hive 集成 Tez。

本章资源下载

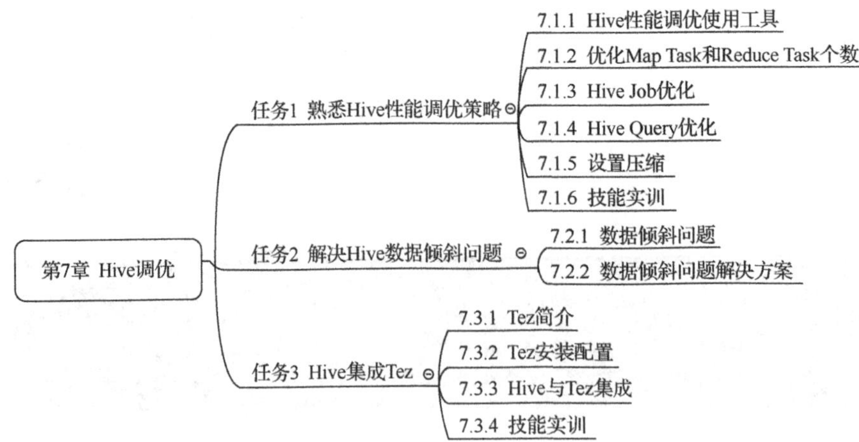

HiveQL 为用户提供了非常简洁的类 SQL 语句，故在大多数情况下，用户不需要了解 Hive 内部复杂的查询、解析、规划、优化和执行等过程。但是在实际工作中，一些低效的查询语句执行会非常耗时，并且会浪费集群资源。因此，随着用户与 Hive 接触越来越多、业务场景越来越复杂以及数据量越来越大，为了提高 Hive 执行任务的效率，用户必须要研究 Hive 性能调优。本章将从不同的方面来介绍 Hive 性能调优。

任务 1 熟悉 Hive 性能调优策略

【任务描述】

了解 Hive 常见的性能调优方式。

【关键步骤】

（1）熟悉 Hive 性能调优使用工具。

（2）熟悉常用的优化方法。

7.1.1 Hive 性能调优使用工具

1. EXPLAIN

HiveQL 是一种声明式语言，用户提交的交互式查询，Hive 会将 HiveQL 语言提交的交互式查询转换成 MapReduce 任务。Hive 提供的 EXPLAIN 命令可显示查询语句的执行计划，因此用户可以通过显示的信息了解 Hive 是如何将查询转换成 MapReduce 任务的。

【语法】

```
EXPLAIN [EXTENDED|DEPENDENCY|AUTHORIZATION] query
```

其中：

- EXTENDED 提供执行计划关于操作的额外信息，比如文件路径；
- DEPENDENCY 提供 JSON 格式的输出，包括查询所依赖的表和分区的列表；
- AUTHORIZATION 提供所有需要授权的实体，包括查询的输入/输出和认证失败。

使用 EXPLAIN 命令可将 Hive 查询转换成阶段（stage）序列（查询本身不会被执行），转换成的 stage 可以是一个 MapReduce 任务，也可以是对元数据（metastore）或文件系统的操作，比如移动或重命名。在默认情况下，Hive 一次只执行一个 stage。

EXPLAIN 的输出包括 3 个部分：

- 查询的抽象语法树（Abstract Syntax Tree）；
- 执行计划中不同 stage 之间的依赖关系；
- 每个 stage 的描述信息，其主要显示操作与数据的对应关系，比如过滤表达式（FilterOperator）、查询表达式（SelectOperator）及文件输出名字（FileSinkOperator）。

理解 Hive 的查询解析和计划并非总是有用的，但是，分析复杂的或执行效率低的查询语句并查看执行计划是非常必要的，特别是在尝试各种调优方式时，用户可以看到调优对执行效率提升所产生的影响。

2. ANALYZE

ANALYZE 关键字可用于搜集表的数值统计信息，以及在执行计划时选择参考。

【语法】

```
ANALYZE TABLE tablename [PARTITION(col1[=val1],col2[=val2],…)]
COMPUTE STATISTICS [noscan]
```

ANALYZE 可统计的字段的含义说明如表 7-1 所示。

表 7-1　ANALYZE 统计字段含义说明

统计字段名称	说明
numPartitions	分区个数
numFiles	文件个数
totalSize	HDFS 存储空间大小
rawDataSize	原始数据（未压缩）大小
numRows	行数

如果指定“noscan”参数，则该命令不会扫描文件数据，仅会统计 numFiles 和 totalSize，统计速度快。如果不指定“noscan”参数，则该命令会扫描文件数据，统计 numRows 和 rawDataSize，同时会将统计的信息更新到元数据表中，下次指定“noscan”参数时还可以得到这些字段。

示例 7-1

获取顾客信息表的统计信息。

关键代码：

```
hive> analyze table customers compute statistics;
```

执行完 ANALYZE 语句后，可以使用“DESCRIBE EXTENDED”命令查看统计的信息。代码如下：

```
DESCRIBE EXTENDED customers;
```

统计结果如下：

```
totalSize=180135315
```

numRows=1902555
rawDataSize=0
numFiles=153

7.1.2 优化 Map Task 和 Reduce Task 个数

1. **设置 Map Task 数量**

在 MapReduce 执行过程中，一个 Job 的 Map Task 数量是由输入分片（InputSplit）决定的，而输入分片的主要决定因素又包括：文件的总个数、文件大小以及集群设置的文件块大小(默认为 128MB)。假设 Hive 表目录下有一个文件 example，其大小为 800MB，HDFS 会将该文件分隔成 7 个块（6 个 128MB 的块和 1 个 32MB 的块），且在执行查询任务时会产生 7 个 Map Task。再假设 Hive 表目录下有 3 个文件：example1、example2 和 example3，它们的大小分别为 15MB、30MB 和 150MB，HDFS 会将它们分隔成 4 个块（15MB、30MB、128MB、22MB），从而即会产生 4 个 Map Task。由此可知，如果文件的大小大于块的大小，则文件会被拆分成多个块。

在 HiveQL 执行过程中，Map Task 的数量过大，不仅会造成 Map 阶段输出文件太小，产生的小文件太多，进而对 HDFS 造成压力，还会增加初始化和创建 Map 的开销。Map Task 的数量太小，又会导致文件处理或查询的并行度低，Job 执行时间过长。因此，Map Task 的数量需要进行合理的控制，我们可以通过以下 2 种方式来控制 Map Task 的数量。

- 减少 Map Task 的个数可以通过合并 Hive 表目录下的文件来实现，这种方式主要针对数据源。
- 增加 Map Task 的个数可以通过增加 Map Task 依赖的前一个 Job 的 Reduce 个数，从而产生更多的文件来实现。

2. **设置 Reduce Task 数量**

（1）Hive 计算 Reduce 的个数

Reduce 个数的设定会极大程度影响任务执行的效率，在不指定 Reduce 个数的情况下，Hive 会根据下面所列的两个属性计算并确定 Reduce 的个数。

- hive.exec.reducers.bytes.per.reducer：每个 Reduce 任务处理的数据量，默认其大小为 1G。
- hive.exec.reducers.max：每个任务对应的最大 Reduce 数，默认是 999。

计算 Reduce 个数的方法为：假设每个任务对应的最大 Reduce 个数为“x”，每个任务处理的数据量为“y”，总的输入数据量为“m”，则可求得 Reduce 的个数 N=min(x,m/y)。

（2）用户调整 Reduce 的个数

用户可以通过改变参数来调整 Reduce 的个数，具体实现方式如下。

- 调整 hive.exec.reducers.bytes.per.reducer 参数的值。
- 重新设置 mapred.reduce.tasks 的值。

Reduce 的个数并非越多越好，因为启动和初始化过多的 Reduce 会耗费太多资源和太长时间。另外，Reduce 个数越多产生的文件也就会越多，如果产生了很多小文件，则会给 HDFS 本身造成非常大的压力。用户应根据实际业务情况对 Reduce 的个数进行适当调整。

示例 7-2

统计顾客信息表中每个姓氏的个数，设置 Reduce 个数为 15。

关键代码：

```
hive> set mapred.reduce.tasks=15;
hive> select customer_lname,count(*) from customers group by customer_lname;
```

代码运行结果如图 7.1 所示。

```
MapReduce Total cumulative CPU time: 1 minutes 16 seconds 640 msec
Ended Job = job_1544060556517_0884
MapReduce Jobs Launched:
Stage-Stage-1: Map: 1  Reduce: 15   Cumulative CPU: 76.64 sec   HD
FS Read: 1243057 HDFS Write: 9189 SUCCESS
Total MapReduce CPU Time Spent: 1 minutes 16 seconds 640 msec
OK
```

图7.1 查看Reduce个数示例代码运行结果

从图 7.1 中可以看出，上述代码执行的 Reduce Task 的个数是 15。

7.1.3 Hive Job 优化

1. **本地模式**（Run In Local Mode）

在大多数情况下，Hadoop Job 须使用分布式执行的方式来处理大数据集，但是当要处理的数据很小时，使用分布式执行的方式处理数据则会增长开销，因为完全分布式模式的启动时间较长，甚至比数据处理时间都长。Hive 可以通过设置属性来将数据处理作业自动转换为本地模式，即使用单台机器处理所有的任务，这对于小数据集处理而言可以明显缩短其执行时间。

本地模式的临时设置方式如下。

```
hive> set hive.exec.mode.local.auto=true;--default false
hive> set hive.exec.mode.local.auto.inputbytes.max=50000000;
hive> set hive.exec.mode.local.auto.input.files.max=5;--default 4
```

本地模式的全局设置需要在 hive-site.xml 中添加属性，方式如下。

```
<property>
    <name>hive.exec.mode.local.auto</name>
    <value>true</value>
</property>
```

需要注意的是，作业必须满足以下条件才能在本地模式下运行。

- 作业的总输入大小必须低于属性“hive.exec.mode.local.auto.inputbytes.max”设置的值。
- Map 任务的总个数必须小于属性“hive.exec.mode.local.suto.input.files.max”设置的值。

➢ Reduce 任务的总个数必须是 1 或者 0。

2. JVM 重用（JVM Reuse）

在默认情况下，Hadoop 会为每一个 Map 和 Reduce 任务分配一个新的 JVM，并且会以并行的方式运行 Map 和 Reduce 任务。但是，当任务是一个只须运行几秒的轻量级作业时，JVM 的启动过程就可能会是一个很大的开销。MapReduce 框架提供了一个选项，即通过重用 JVM 来以串行方式而非并行方式运行 MapReduce Job。设置的方式是在 Hive 的配置文件$HIVE_HOME/conf/hive-site.xml 中配置属性，代码如下：

```
hive> set mapred.job.reuse.jvm.num.tasks=5;
```

该属性的默认值是 1。JVM Reuse 只能在同一个 Job 中的 Map 和 Reduce 任务中起作用，不同 Job 的任务仍然须运行在独立的 JVM 中。

3. 并行执行（Parallel Execution）

Hive 会将一个查询转换成一个或者多个 stage，并按默认顺序执行。在默认情况下，Hive 一次只执行一个 stage。但是，这些 stage 并不总是相互依赖的，相反，它们可以并行执行，这样就可以缩短整个 Job 的执行时间。用户可以通过下列方式设置并行执行。

并行执行的临时设置方式如下。

```
hive> set hive.exec.parallel=true;--default false
hive> set hive.exec.parallel.thread.number=16;--default 8 并行运行的最大数目
```

并行执行的全局设置需要在 hive-site.xml 中配置属性，方式如下。

```
<property>
    <name>hive.exec.parallel</name>
    <value>true</value>
</property>
<property>
    <name>hive.exec.parallel.thread.number</name>
    <value>16</value>
</property>
```

并行执行可以提高集群利用率。但是如果集群利用率已经很高了，则就整体而言并行执行不会有太大的帮助。

4. 推测执行（Speculation Execution）

推测执行是 Hadoop 的一个功能，在分布式集群环境下，程序漏洞、负载不均衡、资源分布不均等原因会造成同一个 Job 的不同 Task 运行速度不一致，某些 Task 的运行速度要明显慢于其他 Task，这些运行速度慢的 Task 会拖慢 Job 的整体执行进度。为了避免这种情况发生，Hadoop 采用了推测执行机制，即首先将执行慢的 Task 加入黑名单，并为这样的 Task 启动一个备份任务，然后让备份任务和原始任务同时处理同一份数据，最后选择最先被成功执行的任务所对应的计算结果为最终结果。

Hadoop 的推测执行功能可以通过设置$HADOOP_HOME/conf/mapred-site.xml 的如下属性实现。

```
<property>
    <name>mapred.map.tasks.speculative.execution</name>
    <value>true</value>
</property>
<property>
    <name>mapred.reduce.tasks.speculative.execution</name>
    <value>true</value>
</property>
```

Hive 框架本身也提供了配置项来控制 Reduce Task 的推测执行，用户可以在 hive 配置文件$HIVE_HOME/conf/hive-site.xml 中配置其属性。

```
<property>
    <name>hive.mapred.reduce.tasks.speculative.execution</name>
    <value>true</value>
</property>
```

注意

推测执行是通过利用更多的资源来提高执行效率的优化策略，因此不建议在资源紧张时使用。

5. **合并小文件**

Hive 实际是在 HDFS 上存储文件，文件数目过多会影响 HDFS 的处理速度，用户可以通过合并 Map 和 Reduce 的结果文件来避免这种影响。

```
<!--是否合并 Map 输出文件，默认为 true -->
<property>
    <name>hive.merge.mapfile</name>
    <value>true</value>
</property>
<!-- 是否合并 Reduce 输出文件，默认为 false -->
<property>
    <name>hive.merge.mapredfiles</name>
    <value>false</value>
</property>
<!-- 合并文件的大小 -->
<property>
    <name>hive.merge.size.per.task</name>
    <value>256*1000*1000</value>
</property>
```

7.1.4 Hive Query 优化

1. **列裁剪**

在利用 HiveQL 查询数据时，并不是所有的任务都需要获取表内所有的数据，有些任务可能只需要读取某些列的数据，比如在查询顾客信息表时，下面的语句只会在表customers 中读取“customer_fname”“customer_lname”“customer_email”“customer_state”列的数据，其他列将会被忽略。

```
select customer_fname,customer_lname,customer_email,customer_state from customers where customer_id=5000;
```

列裁剪的相关属性为 hive.optimize.cp，其默认值是 true。

2. **分区裁剪**

在 Hive 中，可以从不同的维度对表进行分区，并且分区可以嵌套。当需要对目标表的某个分区的数据进行查询时，可以使用分区裁剪。分区裁剪的配置如下。

```
<property>
    <name> hive.optimize.pruner</name>
    <value>true</value>
</property>
```

3. Join

Hive 支持 Join 多表连接查询，如内连接、左外连接、右外连接、全外连接、半连接等。Join 操作的基本原则是：将小表或子查询放在 Join 操作符的左边，因为在执行 Join 操作的 Reduce 阶段时，Join 操作符左边的表会被加载进内存；另外将小表放在 Join 操作符的左边可以减少发生内存溢出错误的概率。

如果一个表足够小，则可以使用 MapJoin 将其整体读入内存中。Join 的操作会在 Map 阶段完成，即在 Map 阶段会直接将另外一张表的数据和内存中表的数据进行匹配，而不需要经过 Shuffle 阶段，这可以在一定程度上节省资源，提高 Join 操作的效率。

在 Hive 0.11 版本之前，必须使用 MapJoin 来启动该优化操作，启动方式如下。

```
select /*+MAPJOIN(smalltable)*/ key,value from smalltable join bigtable on smalltable.key = bigtable.key;
```

从 Hive 0.11 版本开始，Hive 默认启动该优化策略，即在不使用 MapJoin 标记的情况下 Hive 也会在必要的时候触发该操作，将普通的 Join 转换成 MapJoin。用户可以通过下面的属性设置触发该优化策略。

- hive.auto.convert.join：自动使用 MapJoin 优化，其默认值为 true。
- hive.mapjoin.smalltable.filesize：通过该属性测试使用 MapJoin 优化的表的大小，如果表的大小小于该设置值，则其就会被加载进内存中。该设置值的大小默认为 25MB。

示例 7-3

使用 MapJoin 的方式查询顾客的订单信息。

关键代码：

```
hive> select /*+MAPJOIN(customers)*/ customer_fname,customer_lname,order_date
    > from customers join orders on customers.customer_id=orders.order_customer_id;
```

代码运行结果如图 7.2 所示。

```
Launching Job 1 out of 1
Number of reduce tasks is set to 0 since there's no reduce operato
r
Starting Job = job_1544060556517_0886, Tracking URL = http://maste
r:8088/proxy/application_1544060556517_0886/
Kill Command = /opt/hadoop-2.6.0-cdh5.14.2/bin/hadoop job  -kill j
ob_1544060556517_0886
Hadoop job information for Stage-3: number of mappers: 1; number o
f reducers: 0
2019-01-17 13:11:00,026 Stage-3 map = 0%,  reduce = 0%
2019-01-17 13:11:09,307 Stage-3 map = 100%,  reduce = 0%, Cumulati
ve CPU 5.38 sec
MapReduce Total cumulative CPU time: 5 seconds 380 msec
Ended Job = job_1544060556517_0886
MapReduce Jobs Launched:
Stage-Stage-3: Map: 1   Cumulative CPU: 5.38 sec   HDFS Read: 1145
2 HDFS Write: 165 SUCCESS
Total MapReduce CPU Time Spent: 5 seconds 380 msec
OK
```

图7.2　MAPJoin使用示例代码运行结果

从图 7.2 中可以看出，该任务没有 Reduce Task，只有 Map Task，由此可以证明 MapJoin 已生效。

4. GROUP BY 操作

在进行 GROUP BY 操作时，并不是所有的聚合操作都需要在 Reduce 端进行，很多聚合操作都可以先在 Map 端进行部分聚合操作，然后在 Reduce 端得出最终结果。相关属性的配置如下。

```
<!—设定是否在 map 端进行聚合，默认值为 true -->
<property>
    <name>hive.map.aggr </name>
    <value>true</value>
</property>
<!—设定在 map 端进行聚合的条目数，默认值为 100000 -->
<property>
    <name>hive.groupby.mapaggr.checkinterval</name>
    <value>100000</value>
</property>
```

7.1.5　设置压缩

1. 压缩原因

Hive Job 最终会被转换成 MapReduce 任务来执行。MapReduce Job 通常属于 I/O 密集型，即 MapReduce 的性能瓶颈主要在于网络 I/O 和磁盘 I/O，尤其是在数据 Shuffle 的过程中，减少数据量的传输会极大提升 MapReduce 任务的性能。采用数据压缩是减少数据量的一个很好的方式，虽然压缩会消耗 CPU 资源，但是在 Hadoop 集群中，性能瓶颈往往不在于 CPU 所承担的计算压力，因为压缩可以充分利用空闲的 CPU。

2. 常用压缩算法对比

Hive 支持的压缩算法包括 Gzip、Snappy、LZO 和 Bzip2，CDH 版本默认采用的是 Snappy。上述几种压缩算法的对比如表 7-2 所示。

表 7-2　Hadoop 常用压缩算法对比

压缩算法	支持拆分	Hive 自带	压缩率	压缩/解压缩速度
Gzip	否	是	很高	比较快
LZO	是	是	比较高	很快
Snappy	否	是	比较高	很快
Bzip2	是	否	最高	慢

从表 7-2 中可以看出，每一种压缩算法都在压缩/解压缩速度和压缩率间进行了权衡。Bzip2 的压缩率最高，但是其 CPU 消耗也最高。Gzip 的压缩率次之。因此，如果需要提高磁盘空间利用率且减小 I/O 开销，Bzip2 和 Gzip 是不错的选择。LZO 和 Snappy 的压缩率要低于 Bzip2 和 Gzip，但是它们压缩/解压缩的速度要更快，对于经常被读取的数据而言，Snappy 和 LZO 是不错的压缩算法选择。另一个需要考虑的因素就是压缩后的文件是否可以按记录边界进行切分，因为是否可切分直接关系到 MapReduce 任务并行度的高低，每一个切分会被发送到单独的 Map 进程中。Bzip2 和 LZO 提供了块（BLOCK）级别的压缩，每个块中都包含完整的记录信息，如果对任务的并行度有高要求的话，可以考虑使用这两种压缩算法。

各个压缩算法对应的 Class 类如表 7-3 所示。

表 7-3　压缩方式对应 Class 类

压缩算法	Class
Gzip	org.apache.hadoop.io.compress.GzipCodec
LZO	org.apache.hadoop.io.compress.lzo.LzoCodec
Snappy	org.apache.hadoop.io.compress.SnappyCodec
Bzip2	org.apache.hadoop.io.compress.Bzip2Codec

3. **配置压缩**

Hive 提供了 2 种配置压缩方式：中间数据压缩、最终数据压缩。

（1）中间数据压缩

HiveQL 语句最终会被解析成 MapReduce Job。所谓开启 Hive 的中间数据压缩功能，就是在 MapReduce 的 Shuffle 阶段对 Map 端产生的中间结果数据进行压缩。值得注意的是，在 Shuffle 阶段可以选择一个低 CPU 开销的算法。设置“hive.exec.compress.intermediate”的值为“true”可以激活 Hive 中间数据压缩，其默认值是“false”。

示例 7-4

设置 Hive 中间数据压缩。

分析：

首先激活 Hive 中间数据压缩，其次选择合适的压缩算法。该阶段需要使用一个低 CPU 开销的算法，Snappy 压缩算法就是一个比较好的选择，因为其很好地结合了低 CPU

和高压缩执行效率。

关键代码：

```
hive> set hive.exec.compress.intermediate=true;
hive> set mapred.map.output.compression.codec=org.apache.hadoop.io.compress.SnappyCodec;
```

（2）最终数据压缩

通过配置属性“hive.compress.output”可以控制是否对最终输出的内容进行压缩。该属性的默认值为 fasle，因此要想激活这一功能须将该属性设置为“true”。该属性通常会在交互式环境下被临时设置，不建议在 Hive 配置文件中设置全局属性。

示例 7-5

Hive 压缩配置的其他使用方式

设置 Hive 最终数据压缩。

分析：

首先激活最终数据压缩，其次选择合适的压缩算法，此处依旧选择 Snappy 算法。

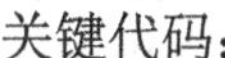

关键代码：

```
hive> set hive.exec.compress.output=true;
hive> set mapred.output.compression.codec=org.apache.hadoop.io.compress.Snappy Codec;
```

Hive 压缩配置的其他使用方式请扫描二维码获取。

7.1.6　技能实训

使用 MapJoin 完成 retail_db 数据库中产品表和订单明细表的连接操作。

要求：

（1）使用 MapJoin 完成；

（2）显示字段包括订单 id、产品价格和产品名称（product_name）。

关键步骤：

（1）设计查询语句，启动 MapJoin；

（2）执行查询，查看运行结果中是否有 Reduce Task 操作。

任务 2　解决 Hive 数据倾斜问题

【任务描述】

了解数据倾斜问题产生的原因及其解决方案。

【关键步骤】

（1）了解数据倾斜问题产生的原因。

（2）学会解决数据倾斜问题。

7.2.1 数据倾斜问题

1. **数据倾斜概述**

数据倾斜是进行大数据计算时经常会遇到的问题之一。数据倾斜是指数据的分布不平衡造成数据大量集中到一点，进而形成数据热点。数据倾斜主要发生在任务进度长时间维持在 99%或者 100%附近，此时少量的 Reduce 子任务未完成，未完成的 Reduce 子任务处理的数据量又与其他 Reduce 子任务差异过大，导致其运行时间远大于其他 Reduce 子任务的平均时长，即这样的 Reduce 子任务成为了整个 Job 的短板，其运行时间直接影响着 Job 的最终运行时间。对该类问题进行优化可以提高 Job 的性能。

2. **产生数据倾斜的原因**

（1）Key 的分布不均匀，某些 Key 太集中。前面说过，Reduce 的数据量差异过大就会造成数据倾斜。Reduce 阶段的数据是由分区来决定的，默认的分区算法是对 Key 求 Hash 值，根据 Hash 值决定该 Key 被分到某个分区，然后再被发送到某个 Reduce。如果 Key 特别集中或者相同，计算得出的 Hash 值一样，就会导致 Key 被全部送到同一个 Reduce，进而就会造成该 Reduce 的数据量过大。

（2）业务数据本身的特性，比如某些业务数据作为 Key 本身就很集中，这也会导致数据倾斜。

（3）建表考虑不周。

（4）某些 HiveQL 语句本身就存在数据倾斜问题。

7.2.2 数据倾斜问题解决方案

1. **参数调节**

```
hive> hive.map.aggr = true
hive> hive.groupby.skewindata=true
```

数据倾斜的时候进行负载均衡，即将以上参数设定为 true 时，生成的查询计划会有两个 MR Job。在第一个 MR Job 中，Map 的输出结果集合会被随机分布到 Reduce 中，每个 Reduce 做部分聚合操作并输出结果，如此处理的结果是相同的 Group By Key 有可能被分发到不同的 Reduce 中，从而达到负载均衡的目的。在此基础上，第二个 MR Job 再根据预处理的数据结果将 Group By Key 分布到 Reduce 中（这个过程可以保证相同的 Group By Key 被分布到同一个 Reduce 中），进而完成最终的聚合操作。

2. **SQL 语句优化**

（1）解决 Join 出现的数据倾斜问题

① 大小表 Join

产生原因：Hive 在进行 Join 时会按照 Join 对 Key 进行分发，而在 Join 左边的表的数据会首先被读入内存，如果左边表的 Key 相对分散，读入内存的数据会比较小，Join 任务执行就会比较快。而如果左边表的 key 比较集中，且这张表的数据量很大，那么数据倾斜问题就会比较严重。

解决方式：使用 MapJoin 让小表先进内存，并在 Map 端完成 Join 操作。关于 MapJoin 的配置及使用方式前面已经介绍过了，此处不再赘述。

② 大表 Join 大表

产生原因：业务数据本身的特性导致 Join 操作的两个表都是大表。比如，假设用户表 users 有 300 万数据，对 users 表和日志数据进行 Join 操作，语句如下。

```
select * from logger log left outer join users u on log.user_id = u.user_id;
```

利用上述语句将 users 表分发到 Map 上是一个非常大的开销，并且这么大的数据量也不适合使用 MapJoin，而如果使用原始的 Join 操作就会产生数据倾斜问题。

解决方式：根据业务进行数据裁剪。当天登录的用户量其实远小于 users 表中存储的用户量，因此可以先只查询当天登录的用户，语句如下。

```
select distinct user_id from logger where login_date='2018-12-28';
```

上述语句可以查询出当天登录的用户 id，该数据量一般不会太大，因此可以对 users 表与查询出来的数据集进行 MapJoin 操作，如此大表 Join 大表的问题就被转化为小表 Join 大表。语句如下。

```
select /*+MAPJOIN(log1)*/u1.* from (select distinct user_id from logger where login_date='2018-12-28' ) log1 join users u1 on log1.user_id = u1.user_id;
```

上述语句使用 MapJoin 的方式可以查询出当天登录的用户的所有信息，并且其数据量不是很大，因此可以使用原始的 logger 表与上述语句进行 MapJoin，进而完成大表与大表连接，并且可以有效避免数据倾斜问题，语句如下。

```
select /*+MAPJOIN(u2)*/* from logger log2
left outer join
(
    select /*+MAPJOIN(log1)*/u1.*
    from (select distinct user_id from logger where login_date='2018-12-28' ) log1
    join users u1 on log1.user_id=u1.user_id
) u2
on log2.user_id = u2.user_id;
```

（2）COUNT DISTINCT 大量相同特殊值

产生原因：使用 COUNT DISTINCT 时，该字段中存在大量值为 NULL 或空的记录。

解决方式：使用 COUNT DISTINCT 时，对值为空的情况进行单独处理。如果是在计算 COUNT DISTINCT，则可以不用处理而是直接过滤，然后在最终结果中加 1。如果还有其他计算，则需要进行 GROUP BY，即首先对值为空的记录进行单独处理，然后将其和其他计算结果进行 UNION。

① 仅计算 COUNT DISTINCT

```
select cast(count(distinct user_id)+1 as bigint) as user_cnt
from user
where user_id is not null and user_id <> '';
```

② 计算完 COUNT DISTINCT 后进行 GROUP BY

在 Hive 中经常会遇到 COUNT DISTINCT 操作，这会导致最终只有一个 Reduce。针对该问题，我们可以先进行 GROUP BY 然后再在其外面嵌套一层 COUNT，例如下列代码所示对日志数据的处理。

```
select day,
count(case when type='session' then 1 else null end) as session_cnt,
count(case when type='user' then 1 else null end) as user_cnt
from (
  select day,type
  from (
    select day,session_id,'session' as type from log
    union all
    select day user_id,'user' as type from log
  )
  group by day,type
)t1
group by day;
```

（3）group by 操作

产生原因：数据中某一个 Key 值大量出现时，HiveQL 语句中进行 group by 操作就会导致对应 Key 的 Reduce 阶段出现数据倾斜。

解决方式：针对上述问题的解决策略是对 Key 进行加盐处理，主体思路是进行两阶段聚合。第一阶段（局部聚合）：首先给每个相同的 Key 加上一个随机数，此时原来相同 Key 就会变成不同的 Key。例如，使用 10 以内的随机数进行加盐处理，则(user,1)、(user,1)、(user,1)、(user,1)就会变成(1_user,1)、(1_user,1)、(2_user,1)、(2_user,1)，然后对加上随机数后的数据进行 SUM 或者 COUNT 局部聚合操作，聚合后的结果为(1_user,2)、(2_user,2)。第二阶段（全局聚合）：首先将 Key 添加的随机数去掉，则可获得(user,2)、(user,2)，然后对数据进行全局聚合操作，就可以得到最终的结果(user,4)。group by 数据倾斜处理过程如图 7.3 所示。这种处理方案对于聚合类的 Shuffle 操作导致的数据倾斜问题具有较好的处理效果，一般情况下均可解决或缓解此类数据倾斜问题。

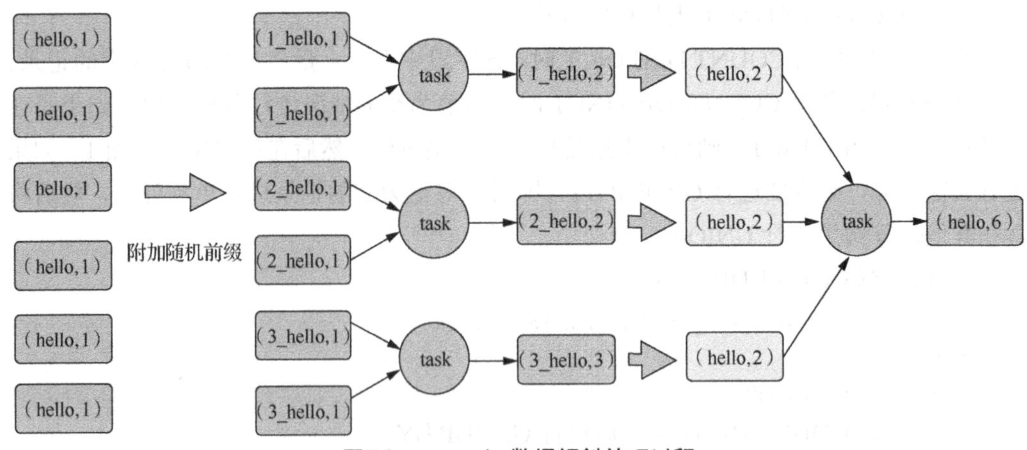

图7.3　group by数据倾斜处理过程

3. **空值产生的数据倾斜**

产生原因：在日志中常会出现信息丢失的情况，比如日志中的 user_id 丢失，此时如果取其中为空值的 user_id 和用户表中的 user_id 进行关联，就会产生数据倾斜的问题。

解决方法 1：user_id 为空的不参与关联。

```
select * from log a join users b on a.user_id is not null and a.user_id = b.user_idunion all select * from log a where a.user_id is null;
```

解决方法 2：赋予空值新的 Key 值。

```
select * from log a left outer join users b on case when a.user_id is null then concat('hive',rand() ) else a.user_id end = b.user_id;
```

结论：方法 2 比方法 1 效率更高。方法 1 中 log 读取两次，Job 数是 2，而方法 2 的 Job 数是 1。这个优化适用于无效 id（比如-99、''、null 等）产生的倾斜问题。把空值的 Key 变成一个字符串加随机数，就能把倾斜的数据分到不同的 Reduce 上，进而即可解决数据倾斜问题。

4. **不同数据类型关联产生数据倾斜**

产生原因：如果用户表中 user_id 字段为 int 类型，log 表中 user_id 字段既有 string 类型也有 int 类型，则在按照 user_id 对两个表进行 Join 操作时，默认的 Hash 操作就会按 int 类型的 user_id 来分配记录，进而即会导致所有 string 类型 user_id 的记录都被分配到一个 Reducer 中。

解决方法：把数字类型转换成字符串类型。代码如下：

```
select * from users a left outer join logs b on a.usr_id = cast(b.user_id as string);
```

任务 3 Hive 集成 Tez

【任务描述】

完成 Hive 与 Tez 的集成。

【关键步骤】

（1）下载安装包。

（2）安装 Tez。

（3）集成 Hive 与 Tez。

（4）启动并验证。

7.3.1 Tez 简介

Tez 是 Apache 开源的支持有向无环图（Directed Acyclic Graph，DAG）作业的计算框架，其可以极大程度地提升 DAG 作业的性能。Hadoop 的 MapReduce 是一个大数据批处理平台，但其并不适用于近乎实时的数据处理场景，如机器学习。创建 Tez 的目的就是使 Hadoop 适用于这些应用场景。

Tez 源于 MapReduce 框架，其核心思想是将 Map 和 Reduce 两个操作进行进一步拆分，即 Map 被拆分成 Input、Processor、Sort、Merge 和 Output，Reduce 被拆分成 Input、Shuffle、Sort、Merge、Processor 和 Output 等，拆分后的元操作可以任意组合，进而就会产生新的操作，这些新的操作经过控制程序组装后可形成一个大的 DAG 作业。

Tez 对外提供 6 种可编程组件，分别介绍如下。

（1）Input：对输入数据源进行抽象，用于解析输入数据的格式，并吐出 Key/Value。

（2）Output：对输出数据源进行抽象，用于将用户程序产生的 Key/Value 写入文件系统。

（3）Partitioner：对数据进行分片，类似于 MR 中的 Partitioner。

（4）Processor：对计算进行抽象，即从 Input 中获取数据并处理，然后通过 Output 输出。

（5）Task：对任务进行抽象，每个 Task 均由 Input、Output 和 Processor 3 部分组成。

（6）Master：管理各个 Task 的依赖关系，并按依赖关系的顺序执行 Task。

Tez 计算框架的引入可以弥补现有 MR 框架在迭代计算和交互式计算方面所存在的不足。Tez 是基于 YARN 的，因此其可以与原有的 MapReduce 共存。Hive 从 0.13 版本开始支持 Tez，选择 Tez 作为 Hive 执行引擎，可以在一定程度上提高 Hive 的执行速度。

7.3.2 Tez 安装配置

1. 安装包下载

请读者自行到官网下载 Tez 安装包。本教材选用的是 0.8.5 版本的源码包。

安装环境要求如下。

操作系统：CentOS 7.5。

Java 环境：Java 8。

Hadoop 环境：hadoop-2.6.0-cdh5.14.2。

Maven 环境：Maven3。

2. 环境准备

（1）安装编辑 Tez 需要的系统包

```
$ yum –y install gcc gcc-c++ libstdc++-devel make build
```

（2）安装 Protobuf2.5.0

① 下载安装包

请读者自行到官网下载 Protobuf 2.5.0 安装包。

② 解压安装

```
$ tar –zxf protobuf-2.5.0.tar.gz
$ cd protobuf-2.5.0/
$ ./configure
$ make & make install
```

③ 验证是否安装成功

```
$ protoc --version
```

验证结果如下。

```
libprotoc 2.5.0
```

3. Tez 编译安装

（1）解压 Tez 源码包

```
$ tar -zxf apache-tez-0.8.5-src.tar.gz
```

（2）修改属性

① 修改源码包中的 pom.xml 文件，并将 hadoop.version 属性设置成对应环境的 Hadoop 版本，配置如下。

```
<properties>
    <hadoop.version>2.6.0-cdh5.14.2<hadoop.version>
</properties>
```

由于 Hadoop 环境是 CDH 版本，所以还需要添加 Cloudera 的 Maven 仓库地址，配置如下。

```
<repositories>
    <repository>
        <id>cloudera</id>
        <url>https://repository.cloudera.com/artifactory/cloudera-repos/</url>
        <name>Cloudera Repositories</name>
        <snapshots>
        <enabled>false</enabled>
        </snapshots>
    </repository>
</repositories>
<pluginRepositories>
    <pluginRepository>
        <id>cloudera</id>
        <name>Cloudera Repositories</name>
        <url>https://repository.cloudera.com/artifactory/cloudera-repos/</url>
    </pluginRepository>
</pluginRepositories>
```

屏蔽 pom.xml 文件中“modules”标签中的 tez-ext-service-tests、tez-ui、tez-ui2 三个模块，若不屏蔽，则无法通过编译。

② 修改 Tez 源码。在解压 Tez 的源码包下修改 tez-mapreduce/src/main/java/org/apache/tez/mapreduce/hadoop/mapreduce/JobContextImpl.java 类，并在 JobContextImpl.java 类的最后添加如下方法。

```
/**
 * Get the boolean value for the property that specifies which classpath
 * takes precedence when tasks are launched. True - user's classes takes
 * precedence. False - system's classes takes precedence.
 * @return true if user's classes should take precedence
```

```
 */
 @Override
public boolean userClassesTakesPrecedence() {
  return getJobConf().getBoolean(MRJobConfig.MAPREDUCE_JOB_USER_CLASSPATH_FIRST, false);
}
```

（3）编译 Tez 源码

```
$ cd apache-tez-0.8.5-src
$ mvn clean package -DskipTests=true -Dmaven.javadoc.skip=true
```

（4）安装 Tez

将编译目录下的 tez-dist/target/tez-0.8.5-minimal 目录上传到 HDFS 中，命令如下。

```
$ cd tez-dist/target
$ hdfs dfs -mkdir -p /app/tez-dir
$ hdfs dfs -put tez-dist/target/tez-0.8.5-minimal /app/tez-dir
$ hdfs dfs -chmod -R 777 /app/tez-dir
```

执行完上述命令即可完成 Tez 的编译安装。Tez 环境搭建视频请扫描二维码观看。

Tez 环境搭建

7.3.3 Hive 与 Tez 集成

前面已经完成了 Tez 的编译安装，本节将介绍如何使用 Hive 集成 Tez。

1．配置文件设置

（1）创建 tez_site.xml 文件

在{HIVE_HOME}/conf 目录下创建 tez-site.xml 文件，内容如下：

```
<configuration>
    <property>
        <name>tez.lib.uris</name>
        <value>${fs.defaultFS}/app/tez-dir/tez-0.8.5-minimal,
               ${fs.defaultFS}/app/tez-dir/tez-0.8.5-minimal/lib</value>
    </property>
    <property>
        <name>tez.use.cluster.hadoop-libs</name>
        <value>true</value>
    </property>
</configuration>
```

（2）修改 Hive 配置文件

在 hive-site.xml 文件中配置如下属性。

```
<property>
    <name>hive.tez.container.size</name>
    <value>1024</value>
 </property>
```

（3）上传 jar 包

将下列三个 jar 包

```
${HADOOP_HOME}/share/hadoop/mapreduce1/hadoop-core-2.6.0-mr1-cdh5.14.2.jar
```

```
${HADOOP_HOME}/share/hadoop/mapreduce/hadoop-mapreduce-client-core-2.6.0-cdh5.8.0.jar
${HADOOP_HOME}/share/hadoop/mapreduce/hadoop-mapreduce-client-common-2.6.0-cdh5.8.0.jar
```

上传到 HDFS 的文件目录/app/tez-dist/target/tez-0.8.5-minimal 下。

2. 在 Hive 中使用 Tez

在 Hive 中使用 Tez，只须在交互式环境中进行如下设置即可。

```
hive> set hive.execution.engine=tez;
```

设置完成后，Hive 的执行引擎就变成了 Tez。

Hive 集成 Tez 的方式请扫描二维码获取。

Hive 集成 Tez

示例 7-6

使用查询验证 Hive 的执行引擎为 Tez。

关键代码：

```
hive> set hive.execution.engine=tez;
hive> select count(*) from empdb.emp;
```

代码运行结果如图 7.4 所示。

```
Query ID = root_20190105145050_89b1a513-828b-483c-9ef2-962026cead30
Total jobs = 1
Launching Job 1 out of 1


Status: Running (Executing on YARN cluster with App id application_1530498791691_0643)

--------------------------------------------------------------------------------
        VERTICES      STATUS  TOTAL  COMPLETED  RUNNING  PENDING  FAILED  KILLED
--------------------------------------------------------------------------------
Map 1 .........   SUCCEEDED     57         57        0        0       0       0
Reducer 2 ......  SUCCEEDED      1          1        0        0       0       0
--------------------------------------------------------------------------------
VERTICES: 02/02  [==========================>>] 100%  ELAPSED TIME: 97.34 s
--------------------------------------------------------------------------------
OK
4
Time taken: 126.385 seconds, Fetched: 1 row(s)
```

图7.4　验证Tez示例代码运行结果

从图 7.4 中可以看出，执行的方式跟 MapReduce 引擎的执行日志明显不一样，但是没有体现 Tez。图 7.4 中提供的“App id”为“application_1530498791691_0643”。可以根据该 App id 查看 Job 的状态，如图 7.5 所示。

```
Application Report :
        Application-Id : application_1530498791691_0643
        Application-Name : HIVE-7412ffea-1765-49dd-902c-797e2ac4223f
        Application-Type : TEZ
        User : root
        Queue : root.root
        Start-Time : 1546671012936
        Finish-Time : 1546671455477
        Progress : 100%
        State : FINISHED
        Final-State : SUCCEEDED
        Tracking-URL : N/A
        RPC Port : 43310
        AM Host : localhost
        Aggregate Resource Allocation : 1137348 MB-seconds, 1106 vcore-seconds
        Log Aggregation Status : SUCCEEDED
        Diagnostics : Session stats:submittedDAGs=0, successfulDAGs=1, failedDAGs=0, killedDAGs=0
```

图7.5　YARN中Job的状态信息

从图 7.5 中可以看出，字段“Application-Type”的值为“TEZ”，由此可以证明 Job 的执行引擎为 Tez，Tez 安装及其与 Hive 集成已经完成。

如果要在 Hive 中使用 MapReduce，则须设置：

```
hive> set hive.execution.engine=mr;
```

7.3.4 技能实训

安装 Tez，并完成 Hive 与 Tez 的集成。

关键步骤：

（1）安装编译 Tez 所需要的系统包；

（2）安装 Protobuf 2.5.0；

（3）下载 Tez 安装包；

（4）修改 Tez 源码及 pom.xml 文件；

（5）编译安装 Tez；

（6）集成 Hive 与 Tez。

本章小结

- 使用 EXPLAIN 命令查询 HiveQL 语句的执行计划可以方便开发人员进行 Hive 性能调优。
- 使用 ANALYZE 命令获取 Hive 表的统计信息可以方便开发人员编写合适的 HiveQL 语句。
- 通过拆分 Hive 表数据对应的文件可以增加 HiveQL 执行过程中的 Map Task 的个数，进而可以增加并行度。
- 通过合并小文件可以减少 Map Task 的个数。
- 合理地设置 Reduce Task 的个数可以达到优化 Hive Job 的效果。
- Hive Job 的优化方式有设置本地模式、JVM 重用、并行执行、推测执行、合并小文件等。
- HiveQL 语句的合理编写可以达到优化 Hive 性能的目的，尤其是在进行 Join 和 Group By 操作时需要注意避免产生数据倾斜。
- Hive 支持多种压缩算法（Gzip、LZO、Snappy、Bzip2），选择合适的压缩算法可以减少 Hive Job 的执行时间。
- 将 Tez 作为 Hive 的执行引擎可以提升 Hive Job 的执行效率。

本章作业

一、简答题

1. 简述 Hive 数据倾斜问题产生的原因。
2. Hive Job 的优化有哪些？
3. Hive 支持的压缩算法有哪些？

二、编码题

1．现有两张表：产品表和日交易表，它们的结构如表 7-4 和表 7-5 所示。

表 7-4 产品表

product_id	product_price	product_name
p_1001	80	T 恤
P_1002	100	裤子
P_1003	1100	冰箱
P_1004	750	衣服
P_1005	680	帽子
…	…	…

表 7-5 日交易表

transaction_id	product_id	transaction_date
T_1001	p_1001	2018-11-11
T_1002	P_1001	2018-11-11
T_1003	P_1002	2018-11-11
T_1004	P_1002	2018-11-11
T_1005	P_1003	2018-11-11
…	…	…

上述两张表的数据量都在 500 万以上，请设计合适的 HiveQL 语句以查询出当日卖出产品的价格及名称。

2．现有在线教育的学生表 students 和日学习记录表 day_study，它们的结构如表 7-6 和表 7-7 所示。

表 7-6 students

stu_id	stu_name	stu_age	stu_sex
S_1001	zhangsan	21	男
S_1002	lisi	20	女
S_1003	wangwu	19	男
S_1004	zhaoliu	18	女
…	…	…	…

表 7-7 day_study

study_id	stu_id	course_id	study_date
1	S_1001	1	2018-11-11
2	S_1001	2	2018-11-11
3	S_1001	3	2018-11-11

续表

study_id	stu_id	course_id	study_date
4	S_1002	4	2018-11-11
...	...	...	

其中，students 的数据量在 10 万左右，day_study 的数据量在 500 万以上，请编写 HiveQL 语句以查询出每条学习记录对应的学生信息：姓名、年龄、性别。

3．使用 ANALYZE 命令分析并统计“retail_db”数据库中的数据信息。

第 8 章

Hive 与 HBase 集成

技能目标

➢ 理解 Hive 整合 HBase 的原理。
➢ 掌握 Hive 与 HBase 的集成方法。

本章任务

任务 1 理解 Hive 与 HBase 集成的场景及原理。
任务 2 实现 Hive 与 HBase 集成。
任务 3 使用 Phoenix 操作 HBase 数据库。

本章资源下载

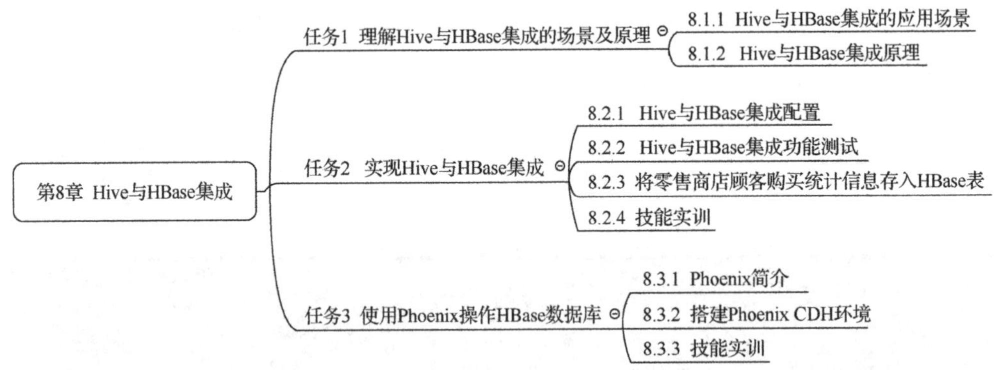

HBase 是 Hadoop 生态系统中的 NoSQL 数据库，用于快速存储和访问海量数据。HBase 的缺点是语法格式比较底层，没有类 SQL 的语法，在实际应用中操作和计算都非常不方便。Hive 作为建立在 Hadoop 之上的数据仓库工具，其提供了类 SQL 的语法，极大地简化了开发工作。所以将两者结合起来使用，通过 Hive 这个客户端对 HBase 的数据进行操作和查询，可以简化 HBase 原生的复杂 API 操作。本章主要对 Hive 与 HBase 的集成进行讲解。

任务 1 理解 Hive 与 HBase 集成的场景及原理

【任务描述】

了解 Hive 与 HBase 集成的应用场景，理解 Hive 与 HBase 集成的原理。

【关键步骤】

（1）了解 Hive 与 HBase 集成的应用场景。

（2）理解 Hive 与 HBase 集成的原理。

8.1.1 Hive 与 HBase 集成的应用场景

Hive 与 HBase 都是基于 Hadoop 的技术，Hive 是类 SQL 的执行引擎，HBase 可以提供实时交互式查询，两种技术各有特点：Hive 具有类 SQL 的执行优势，而 HBase 具有交互式的特点。两者结合使用可以充分发挥两者各自的特点。

1．将 Hive ETL 操作数据存入 HBase

Hive 与 HBase 集成的第一种应用场景是将 Hive ETL 操作数据装载到 HBase 表中，供前端进行快速的查询与展示，数据源可以是 Hive 表或者文件，这种架构（如图 8.1 所示）解决了 Hive 查询慢的问题。

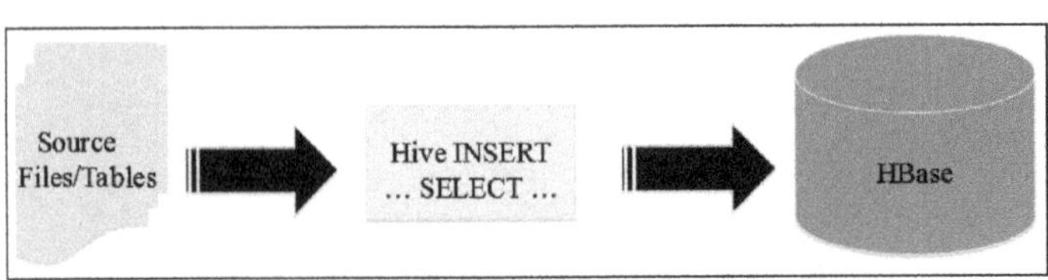

图8.1　Hive ETL操作数据存入HBase架构

2．HBase 作为 Hive 的数据源

Hive 与 HBase 集成后的第二种应用场景是 Hive 作为客户端工具，让 HBase 支持类 SQL 的操作，比如可以执行 JOIN、GROUP BY 等 SQL 查询语法。这种应用场景将 HBase 作为 Hive 数据仓库的数据源，可以解决 HBase 不适合做复杂数据分析的问题，其架构如图 8.2 所示。

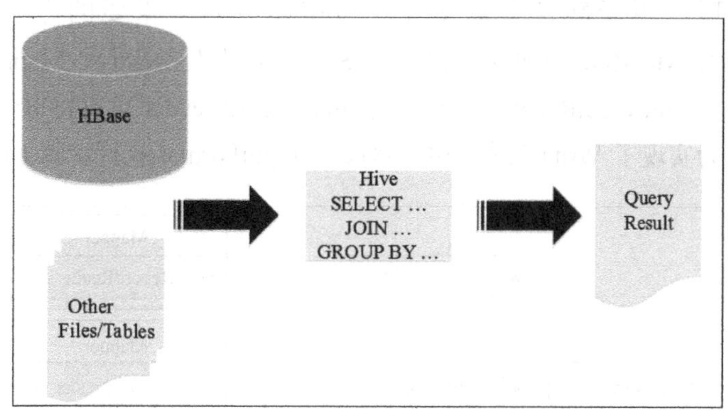

图8.2　HBase作为Hive数据源架构

3．构建低延时的数据仓库

Hive 与 HBase 整合后的第三种应用场景是构建低延时的数据仓库，利用 HBase 的快速读写能力可完成数据实时查询，也可以使用 Hive 查询 HBase 中的数据以完成复杂的数据分析。低延时数据仓库架构如图 8.3 所示。

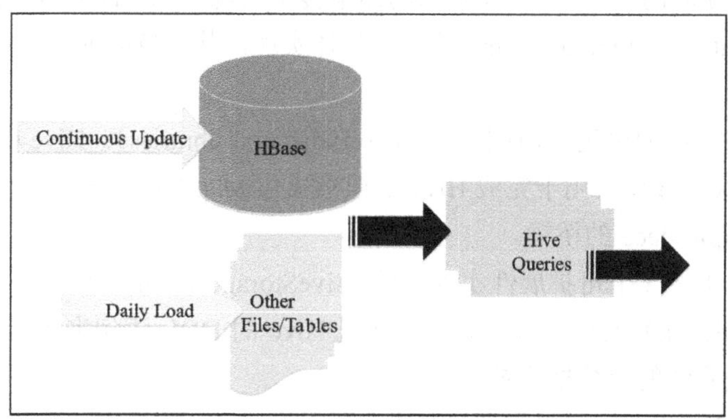

图8.3　低延时数据仓库架构

8.1.2　Hive 与 HBase 集成原理

1．Hive 存储处理器（HiveStorageHandler）

Hive 提供了存储处理器（Storage Handler）以方便用户使用 SQL 读写外部数据源。直接用原来的方法操作 ElasticSearch、Kafka、HBase、Cassandra 等数据源，对开发人员来说有一定的门槛，而借助 Storage Handler，开发人员可以很方便地编写简单且熟悉的类 SQL 语句以访问上述数据源。Hive 框架提供了 HiveStorageHandler 接口，用户可以使

用 Java 实现通过该接口添加新的存储器，比如与 HBase 集成的 HBaseStorageHandler 就是该接口的实现类。

Hive 存储处理器功能的实现基于了 Hive 及 Hadoop 的扩展性。

（1）输入格式化（InputFormat）：在 Hadoop 中，InputFormat 接口可以将来自不同数据源的数据的格式转换成 MapReduce Job 输入的格式，其中 TextInputFormat 就是 InputFormat 的一个具体实现，用于转换文本文件。该接口有两个重要的方法：getSplits()和 getRecordReader()。MapReduce 框架可根据 getSplits()方法生成若干个 Mapper。一个 HDFS 文件可能会有多个 Split，getRecordReader()返回的 RecordReader 负责将每一个 Split 对应的二进制数据转换为实现了 Writable 接口的类对象。InputFormat 执行步骤如图 8.4 所示。

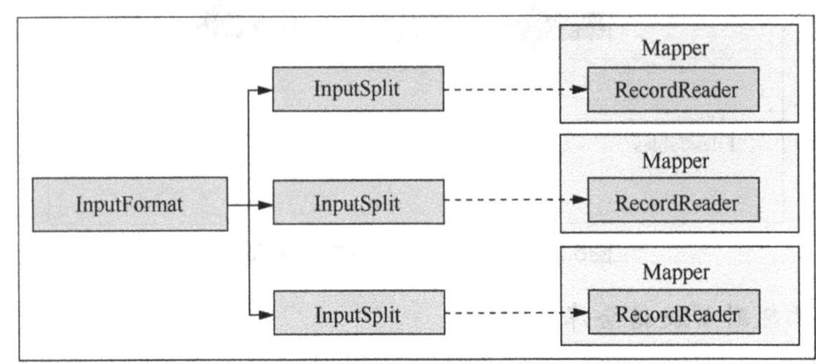

图8.4 InputFormat执行步骤

（2）输出格式化（OutputFormat）：Hadoop 同时还提供了 OutputFormat 接口，该接口会获取到一个 Job 的输出，然后将输出的结果传至实现了 Writable 接口的实体对象中。TextOutputFormat 就是 OutputFormat 的一个具体实现，其可持续地将结果输出到 HDFS 上的文本文件中。

（3）序列化/反序列化包（serialization/deserialization libraries）：Hive 中提供了序列化/反序列化（SerDe）接口，用于完成 Hive 数据格式与 MapReduce 数据格式之间的转换，它是 Hive 与 MapReduce 的桥梁。

除了依赖于上述 3 个可扩展性实现以外，HiveStorageHandler 还需要实现新的元数据钩子（HiveMetaHook）接口，该接口允许使用 Hive 的 DDL 语句来定义和管理 Hive 的元数据以及其他系统的元数据目录。

HiveStorageHandler 接口中定义的方法介绍如下。

- public Class<? extends InputFormat> getInputFotmatClass()：该方法返回 InputFormat 接口的一个实现类。
- public Class<? extends OutputFormat> getOutputFormatClass()：该方法返回 OutputFormat 接口的一个实现类。
- public Class<? extends SerDe> getSerDeClass()：该方法返回 SerDe 接口的一个实现类。
- public HiveMetaHook getMetaHook()：该方法返回元数据钩子的一个实现类。
- public void configureTableJobProperties(TableDesc tableDesc,Map<String,String>

jobProperties)：该方法的作用主要是允许 HiveStorageHandler 设置需要绑定的属性，以供 JobContext.getConfiguration()获取对应的属性。

具体的执行过程如下。

当执行查询语句“select customer_fname from customers;”时，假设 customers 是使用 HiveStorageHandler 的实现类创建的，则执行 HiveQL 查询时，Hive 的 SQL 执行引擎会将该查询转换为一个 MapReduce 应用。该应用会包含若干个 Mapper，Mapper 的个数取决于 HiveStorageHandler 实现类中 getInputFormatClass()方法返回的 InputFormat 对象的 getSplits()方法。Mapper 内部的<Key,Value>则由返回的 InputFormat 对象的 getRecordReader()方法获得的 RecordReader 对象来生成，其中 Key 和 Value 都实现了 Writable 接口。RecordReader 对象将数据写入到<Key,Value>对象，最后由 SerDe 接口实现类的 deserialize()方法将<Key,Value>对象反序列化成 Hive 的行数据。

2. HBaseStorageHandler

前面已经介绍过 HiveStorageHandler 接口，要实现外部数据源与 Hive 的集成，都要实现该接口中的方法。实现 Hive 与 HBase 集成的类 HBaseStorageHandler 就是该接口的一个实现。实现 Hive 与 HBase 集成主要是利用 Hive 与 HBase 本身对外提供的 API 接口的互相通信，通信原理如图 8.5 所示。

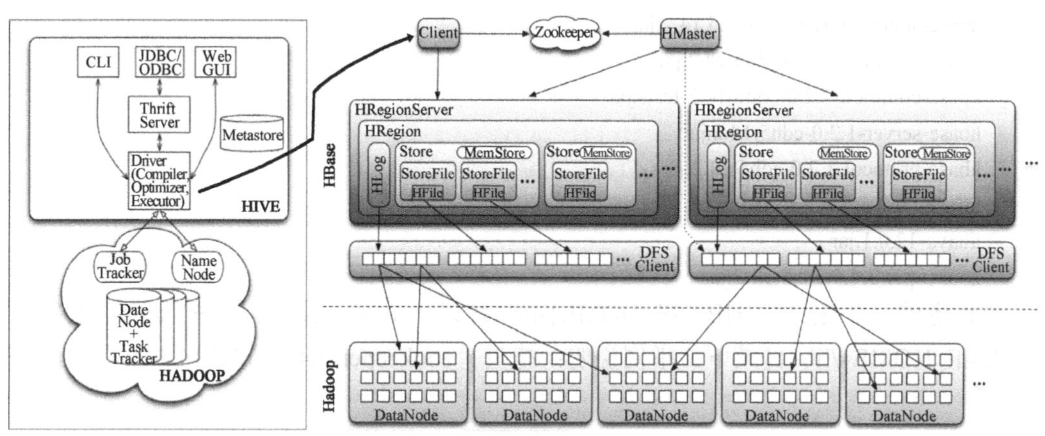

图8.5 Hive与HBase通信原理

API 接口的互相通信主要依靠在${HIVE_HOME}/lib 目录中的 hive-hbase-handler-*.jar 工具类。

HBaseStorageHandler 对 HiveStorageHandler 接口的实现方法如下。

利用 Jar 包中的 HBaseStorageHandler，Hive 可以获取到 Hive 表对应的 HBase 表名、列族以及列。Hive 访问 HBase 数据主要是通过 MapReduce 读取 HBase 表数据，实现方式是使用 HiveHBaseTableInputFormat 切分 HBase 表，并获取 RecordReader 对象以读取数据。读取 HBase 表数据就是使用 HBase API 来构建 Scanner 对象，然后对表进行扫描。如果存在过滤条件，则将 Scanner 对象转换成对应的 Filter 对象。当过滤条件为 RowKey 时，将过滤转换为对 RowKey 的过滤。

任务 2 实现 Hive 与 HBase 集成

【任务描述】

配置并验证 Hive 与 HBase 的集成环境。

【关键步骤】

（1）配置 Hive 与 HBase 的集成环境。

（2）在 HBase 中建立表，并使用 Hive 访问 HBase 表。

（3）将 Hive ETL 数据存入 HBase 表中。

8.2.1 Hive 与 HBase 集成配置

1. 整合 Jar 包

Hive 与 HBase 集成使用的存储处理程序是作为一个独立的模块 hive-hbase-handler-*.jar 来构建的，该模块依赖于 HBase、Guava 和 ZooKeeper 的部分 Jar 包，其名称罗列如下。

```
hbase-client-1.2.0-cdh5.14.2.jar
hbase-annotations-1.2.0-cdh5.14.2.jar
hbase-common-1.2.0-cdh5.14.2.jar
hbase-common-1.2.0-cdh5.14.2-tests.jar
hbase-server-1.2.0-cdh5.14.2.jar
hbase-protocol-1.2.0-cdh5.14.2.jar
htrace-core-3.2.0-incubating.jar
guava-12.0.1.jar
zookeeper-3.4.5-cdh5.14.2.jar
```

上述的 Jar 包在${HBASE_HOME}/lib 目录下，我们须将这些 Jar 包拷贝到${HIVE_HOME}/lib 目录下。举例说明：拷贝“zookeeper-3.4.5-cdh5.14.2.jar”的命令如下。

```
cp ${HBASE_HOME}/lib/ zookeeper-3.4.5-cdh5.14.2.jar ${HIVE_HOME}/lib/
```

其他的 Jar 包可参考上述方法完成拷贝。

2. 修改 Hive 配置文件

（1）配置“hive.aux.jar.path”属性

上一节已经将有所依赖的 Jar 包拷贝到了${HIVE_HOME}/lib 目录下，接下来需要在${HIVE_HOME}/conf/hive-site.xml 配置文件中将 Jar 包的路径配置在“hive.aux.jar.path”属性中，同时也需要将${HIVE_HOME}/lib 目录下的“hive-hbase-handler-1.1.0-cdh5.14.2.jar”配置到该属性中。配置方式如下。

```
<property>
    <name>hive.aux.jars.path</name>
    <!--各 Jar 包之间用逗号分隔 -->
    <value>
```

```
            file:///opt/hive-1.1.0-cdh5.14.2/lib/hive-hbase-handler-1.1.0-cdh5.14.2.jar,
            file:///opt/ hive-1.1.0-cdh5.14.2/lib/zookeeper-3.4.5-cdh5.14.2.jar,
            file:///opt/ hive-1.1.0-cdh5.14.2/lib/hbase-annotations-1.2.0-cdh5.14.2.jar,
            ...
            file:///opt/ hive-1.1.0-cdh5.14.2/lib/ guava-12.0.1.jar
        </value>
    </property>
```

（2）配置“hive.zookeeper.quorum”属性

由于在 HBase 架构中是通过 ZooKeeper 与 HBase 节点进行通信的，因此在${HIVE_HOME}/conf/hive-site.xml 文件中还需要配置“hbase.zookeeper.quorum”属性，并设置 ZooKeeper 节点的主机名。配置方式如下。

```
<property>
    <name>hbase.zookeeper.quorum</name>
    <!--master,slave1,slave2 是 ZooKeeper 安装节点的主机名 -->
    <value>master,slave1,slave2</value>
</property>
```

（3）在 Hive-env.sh 文件中添加 Hadoop 和 HBase 的安装目录路径

在${HIVE_HOME}/conf/hive-env.sh 文件中添加 HADOOP_HOME 和 HBASE_HOME。添加方式如下。

```
export HADOOP_HOME=/opt/hadoop-2.6.0-cdh5.14.2/
export HBASE_HOME=/opt/hbase-1.2.0-cdh5.14.2/
```

上述三个步骤完成后，Hive 与 HBase 的集成配置也就完成了。

8.2.2 Hive 与 HBase 集成功能测试

在讲解 Hive 与 HBase 集成功能测试之前，读者需要了解下列术语。

Hive 自身相关的概念。

➢ 内部表（Managed Table）：被 Hive 管理的表，其元数据由 Hive 管理，且存储在 Hive 的体系里面。

➢ 外部表（External Table）：表的定义由外部的元数据目录所管理，其数据存储在外部系统中。

Hive 存储处理器相关的概念。

➢ 非本地（Non-Native）表：需要通过存储处理器才能管理和访问的表。

➢ 本地（Native）表：Hive 不需要借助存储处理器就可以直接管理和访问的表。

上述 4 种术语进行交叉，即可获得下面 4 种形式的概念定义。

➢ 被管理的本地（Managed Native）表：通过 CREATE TABLE 创建的表。

➢ 外部本地（External Native）表：由 CREATE EXTERNAL TABLE 创建，但是没有 STORED BY 子句。

➢ 被管理的非本地（Managed Non-Native）表：由 CREATE TABLE 创建，同时有 STORED BY 子句。Hive 会在元数据中存储该表的定义，但是不会创建任何文件，hive

存储处理器会向存储数据的系统发出一个请求以创建一个与该表一致的对象结构。

➢ 外部非本地（External Non-Native）表：由 CREATE EXTERNAL TABLE 创建，并且带有 STORED BY 子句。Hive 会在自己的元数据中注册该表的定义信息，并且通过调用存储处理器可以检查这些注册在 Hive 中的信息是否与其他系统中原有的定义信息一致。

1. 创建被 Hive 管理的 HBase 表

在前面的章节中已经介绍过创建表的语法，如下。

【语法】

```
CREATE [EXTERNAL] TABLE [IF NOT EXISTS] table_name
   [(col_name data_type [COMMENT col_comment], …)]
   [COMMENT table_comment]
   [PARTITIONED BY (col_name data_type [col_comment], col_name data_type [COMMENT col_comment], …)]
   [CLUSTERED BY (col_name, col_name, …) [SORTED BY (col_name, …)] INTO num_buckets BUCKETS]
   [
    [ROW FORMAT row_format] [STORED AS file_format]
    | STORED BY 'storage.handler.class.name' [WITH SERDEPROPERTIES (…)]
   ]
   [LOCATION hdfs_path]
   [AS select_statement]
```

一个可以被 Hive 管理的 HBase 表的创建，可以通过在 CREATE TABLE 后面使用 STORED BY 子句来实现。通过新的 STORED BY 子句创建表时，存储处理程序将与该表进行关联。从上述创建表的语句中也可以发现，在 STORED BY 子句后面可以指定存储处理器的类名。STORED BY 子句是 ROW FORMAT 和 STORED AS 子句的替代项，也就是说指定了 STORED BY 子句，则不能再指定 ROW FORMAT 和 STORED AS 子句。可将 WITH SERDEPROPERTIES 作为 STORED BY 子句的一部分，用于指定一些属性，这些属性会被传递给存储处理程序提供的 SerDe。

示例 8-1

使用 HiveQL 创建一个指向 HBase 的 Hive 内部表。

关键代码：

```
create table hbase_table (key int, value string)
stored by 'org.apache.hadoop.hive.hbase.HBaseStorageHandler'
with serdeproperties ("hbase.columns.mapping" = ":key,cf1:val")
TBLPROPERTIES ("hbase.table.name" = "hive_table", "hbase.mapred.output.outputtable" = "hive_table");
```

其中：

➢ hbase.columns.mapping 属性是必需的，用于声明列族和列名，并建立 Hive 表字段与 HBase 中列族/列的映射关系。

➢ hbase.table.name 属性是可选的，可用于指定创建的表在 HBase 中显示的表名。

该属性允许 Hive 和 HBase 设置不同的表名，比如在该示例中，创建的表在 Hive 中的名字是 hbase_table，而在 HBase 中的名字为 hive_table；如果不指定该属性，在 Hive 和 HBase 中显示的表名将完全一样。

➢ hbase.mapred.output.outputtable 属性是可选的，用于指定插入数据时写入的表，如果需要向该表插入数据就需要设置该属性。

执行完上面的语句后，Hive 中会创建一个空表，同时 HBase 数据库中也会存在一个名为“hive_table”的表，并且该表的详细信息可被查看，如图 8.6 所示。

```
Table hive_table is ENABLED
hive_table
COLUMN FAMILIES DESCRIPTION
{NAME => 'cf1', BLOOMFILTER => 'ROW', VERSIONS => '1', IN_MEMORY => 'false', KEEP_DELET
ED_CELLS => 'FALSE', DATA_BLOCK_ENCODING => 'NONE', TTL => 'FOREVER', COMPRESSION => 'N
ONE', MIN_VERSIONS => '0', BLOCKCACHE => 'true', BLOCKSIZE => '65536', REPLICATION_SCOP
E => '0'}
1 row(s) in 0.0290 seconds
```

图8.6 Hive映射HBase表的详细信息

示例 8-2

在示例 8-1 创建的表中插入数据，并在 HBase Shell 中查看插入的数据。

分析：

该表可以是 Hive 中已经存在的一个表，这里选择“retail_db”数据库中的“products”表，选取的列为“product_id”和“product_name”。

关键代码：

```
insert overwrite table hbase_table select product_id,product_name
from retail_db.products limit 5;
```

上面的语句执行完后，HBase 表中查看到的数据结果如图 8.7 所示。

```
ROW                          COLUMN+CELL
 1                           column=cf1:val, timestamp=1547108196161, value=Quest Q64 10 FT. x 10 FT. Slant Leg Instant U
 2                           column=cf1:val, timestamp=1547108196161, value=Under Armour Men's Highlight MC Football Clea
 3                           column=cf1:val, timestamp=1547108196161, value=Under Armour Men's Renegade D Mid Football Cl
 4                           column=cf1:val, timestamp=1547108196161, value=Under Armour Men's Renegade D Mid Football Cl
 5                           column=cf1:val, timestamp=1547108196161, value=Riddell Youth Revolution Speed Custom Footbal
5 row(s) in 0.0200 seconds
```

图8.7 HBase Shell查看Hive映射表的数据

2. 创建 Hive 外部表以映射 HBase 中已经存在的表

创建外部表适用于该表已经存在于 HBase 数据库中，但是 Hive 中并没有与其相关的信息，此时可以通过创建 Hive 外部表来实现 HBase 现有表的类 SQL 查询。

示例 8-3

基于 HBase 数据库中的表创建 Hive 外部表映射。

分析如下。

（1）在建立 Hive 表之前，需要先在 HBase 数据库中建立表，本示例将创建学生成绩“student_scores”表，列族为“stuinfo”。建表语句如下。

```
create 'student_scores',{NAME => 'stuinfo'}
```

（2）在创建的 HBase 表“student_scores”中插入数据。插入数据语句如下。

```
put 'student_scores', 'math', 'stuinfo:score', '99'
put 'student_scores', 'english', 'stuinfo:score', '96'
put 'student_scores', 'chinese', 'stuinfo:score', '86'
```

（3）建立 Hive 表并映射 HBase 表，建立 Hive 字段与 HBase 列族/列的映射。

关键代码：

```
create external table student_scores_hive(course_name string,score int)
stored by 'org.apache.hadoop.hive.hbase.HBaseStorageHandler'
with serdeproperties ("hbase.columns.mapping" = ":key,stuinfo:score")
tblproperties (
"hbase.table.name" = "student_scores",
"hbase.mapred.output.outputtable" = "student_scores"
);
```

执行完上述语句后，可以在 Hive 中查询表“student_scores_hive”的数据，如图 8.8 所示。

```
+----------------------------------+---------------------------+
| student_scores_hive.course_name  | student_scores_hive.score |
+----------------------------------+---------------------------+
| chinese                          | 86                        |
| english                          | 96                        |
| math                             | 99                        |
+----------------------------------+---------------------------+
```

图8.8 student_scores_hive表数据

3. 列映射规则（Column Mapping）

Hive 与 HBase 的字段映射规则由两个 SERDEPROPERTIES 属性来决定，它们分别为 hbase.columns.mapping 属性和 hbase.table.default.storage.type 属性，其中第二个属性从 Hive 0.9 版本开始才可以使用，它的值可以是 string 或 binary，默认值是 string。

目前，可用的列映射规则有些繁琐且具有局限性，使用时需要注意以下 7 点。

（1）在 Hive 表中或者在 SELECT 子句查询的数据集中的每个列，都需要在参数 hbase.columns.mapping 中指定一个对应的条目，即如果某个表有 n 个列，则参数 hbase.columns.mapping 的值中就有 n 个以逗号分隔的条目（各列对应的条目通过逗号进行分隔）。

注意

hbase.columns.mapping 的值中不能出现空格。

（2）匹配条目的形式如下：

```
:key,:timestamp|column-family-name:[column-name][#(binary|string)]
```

其中：

➤ 前面带有“#”标识的是在 Hive0.9 版本中才开始添加的，早期的 Hive 版本会把所有数据的类型都当作 string 类型；

➢ 如果没有给列或者列族指定数据类型，属性 hbase.table.default.storage.type 的值会被当作这些列或者列族的类型，默认值为 string；

➢ 表示类型值（binary|string）的任何前缀都可以用来表示对应的类型，比如#b 代表#binary，#s 代表 string；

➢ 如果将列的类型设置为二进制（binary），则相应的 HBase 单元格的数据应采用 HBase 的 Bytes 类生成。

（3）“:key”映射只能有一个。

（4）在 Hive 0.6 版本以前，将第一个条目作为关键字（key）字段，从 Hive 0.6 版本开始，需要直接通过“:key”的方式来指定。

（5）如果 column-name 没有被指定，Hive 列将映射对应的 HBase 列族的所有列。

（6）在 HBase 1.1 版本后，可以使用“:timestamp”来访问 HBase 时间戳属性，其对应的值的属性只能是 bigint 或者 timestamp。

（7）在实际应用中，有时不需要对 HBase 的所有列族都进行映射，但是没有被映射的列族将不能通过 Hive 表访问其数据；有时也可以把多个 Hive 表映射到同一个 HBase 的表。

4. 多列和多列族映射

在实际应用中，Hive 与 HBase 通常都是多个列映射的。

示例 8-4

将 Hive 中“retail_db”数据库的顾客信息表“customers”中姓氏为“Jones”的数据映射到 HBase 表的两个列族（basicInfo 和 address），其中列族 basicInfo 映射的列为顾客的姓名（customer_fname+customer_lname）和邮件（customer_email），列族 address 映射的列为顾客所在国家（customer_state）、城市（customer_city）和街道（customer_street）。

分析：

➢ 由于 customers 表为外部本地表，因此需要创建新的外部非本地表来存储姓氏为“Jones”的数据；

➢ 在本示例中，Hive 中的多列映射到了 HBase 的两个列族，鉴于 HBase RowKey 的唯一性，须将 customers 表中的“customer_id”作为“:key”的映射；

➢ customers 表中的 customer_fname 和 customer_lname 需要被拼接，以组成顾客的姓名字段。

关键代码：

（1）在 Hive 中创建表

```
create table customers_hive(id string,name string,email string,state string,city string,street string)
stored by 'org.apache.hadoop.hive.hbase.HBaseStorageHandler'
with serdeproperties (
"hbase.columns.mapping" =
":key,basicInfo:name,basicInfo:email,address:state,address:city,address:street"
)
tblproperties (
```

```
"hbase.table.name" = "customer_hbase",
"hbase.mapred.output.outputtable" = "customer_hbase");
```

（2）在表中插入数据

```
insert overwrite table customers_hive
select customer_id,concat(customer_fname,' ',customer_lname) as customer_name,
customer_email,customer_state,customer_city,customer_street
from retail_db.customers where customer_lname='Jones';
```

运行上述代码后，可在 HBase 中查看对应的表数据信息如图 8.9 所示。

```
hbase(main):005:0> scan 'customer_hbase'
ROW                                  COLUMN+CELL
 10078                               column=address:city, timestamp=1547176454275, value=Manati
 10078                               column=address:state, timestamp=1547176454275, value=PR
 10078                               column=address:street, timestamp=1547176454275, value=5282 Silent Landing
 10078                               column=basicInfo:email, timestamp=1547176454275, value=XXXXXXXXX
 10078                               column=basicInfo:name, timestamp=1547176454275, value=Mary Jones
 1009                                column=address:city, timestamp=1547176454275, value=Eagle Pass
 1009                                column=address:state, timestamp=1547176454275, value=TX
 1009                                column=address:street, timestamp=1547176454275, value=4396 Little Pond Crescent
 1009                                column=basicInfo:email, timestamp=1547176454275, value=XXXXXXXXX
 1009                                column=basicInfo:name, timestamp=1547176454275, value=John Jones
 1032                                column=address:city, timestamp=1547176454275, value=Caguas
 1032                                column=address:state, timestamp=1547176454275, value=PR
 1032                                column=address:street, timestamp=1547176454275, value=757 Misty Goose Impasse
```

图8.9　customer_hbase表数据信息

示例 8-5

将 HBase 数据库中已经存在的表 gamer 中多个列族（personalInfo、recordInfo、assetsInfo）的数据映射到 Hive 表中。其中 HBase 中表 gamer 的建表语句如下。

```
create 'gamer',{NAME => 'personalInfo', VERSIONS => 1},{NAME => 'recordInfo', VERSIONS =>
1},{NAME => 'assetsInfo', VERSIONS => 1}
```

执行以下语句以在 gamer 表中插入数据。

```
put 'gamer','row-0001','personalInfo:nickname','QGhappy.Snow'
put 'gamer','row-0001','recordInfo:ranking','one'
put 'gamer','row-0001','assetsInfo:integral','10000'
put 'gamer','row-0001','personalInfo:gemeID','00000000'
put 'gamer','row-0002','personalInfo:nickname','XQMaster'
put 'gamer','row-0002','recordInfo:ranking','two'
put 'gamer','row-0002','assetsInfo: integral','10000'
put 'gamer','row-0002','personalInfo:gemeID','11111111'
```

gamer 表数据信息如图 8.10 所示。

```
hbase(main):015:0> scan 'gamer'
ROW                                  COLUMN+CELL
 row-0001                            column=assetsInfo: integral, timestamp=1547176949069, value=10000
 row-0001                            column=personalInfo:gemeID, timestamp=1547176949106, value=00000000
 row-0001                            column=personalInfo:nickname, timestamp=1547176948978, value=QGhappy.Snow
 row-0001                            column=recordInfo:ranking, timestamp=1547176949026, value=one
 row-0002                            column=assetsInfo: integral, timestamp=1547176949181, value=10000
 row-0002                            column=personalInfo:gemeID, timestamp=1547176951170, value=11111111
 row-0002                            column=personalInfo:nickname, timestamp=1547176949130, value=XQMaster
 row-0002                            column=recordInfo:ranking, timestamp=1547176949154, value=two
2 row(s) in 0.0170 seconds
```

图8.10　gamer表数据信息

分析：

gamer 表中有多个列族，因此在创建 Hive 映射表时需要指定多个列与之对应。

关键代码：

```
create external table gamer_hive(id string,gameID string,nick_name string,ranking string,integral string)
stored by 'org.apache.hadoop.hive.hbase.HBaseStorageHandler'
with serdeproperties (
"hbase.columns.mapping" =
":key,personalInfo:gemeID,personalInfo:nickname,recordInfo:ranking,assetsInfo:integral")
tblproperties ("hbase.table.name" = "gamer", "hbase.mapred.output.outputtable" = "gamer");
```

运行结果：

创建完表后，在 Hive 中查询表数据信息如图 8.11 所示。

gamer_hive.id	gamer_hive.gameid	gamer_hive.nick_name	gamer_hive.ranking	gamer_hive.integral
row-0001	00000000	QGhappy.Snow	one	NULL
row-0002	11111111	XQMaster	two	NULL

图8.11　gamer_hive表数据信息

5. Hive MAP 与 HBase 列族的映射

在前面章节中已经介绍过，Hive 支持 MAP 数据类型，本节将介绍 Hive 的 MAP 数据类型如何与 HBase 列族进行映射。

示例 8-6

将“empdb”数据库中的 emp 表中的“skills_score”和“name”列映射为 HBase 表。

关键代码：

（1）创建内部表

```
create table emp_map(skills_score map<string,int>, name string)
stored by 'org.apache.hadoop.hive.hbase.HBaseStorageHandler'
with serdeproperties (
"hbase.columns.mapping" = "info:,:key"
);
```

其中，“info”表示列族，冒号后可为空，它与 Hive 表的列 skills_score 对应，即 skills_score 的 Key 为 info 列族的列名。

注意

当 hbase.columns.mapping 中有“:key,cf:”这样的值（即列族冒号后面为空）时，Hive 中对应的类型应为 MAP，如果不是 MAP，则创建表将会失败。

（2）向表中插入数据

```
insert overwrite table emp_map select skills_score,name from empdb.emp;
```

运行结果：

数据被插入后，在 HBase 中查询表数据信息如图 8.12 所示。

```
hbase(main):020:0> scan 'empdb.emp_map'
ROW                                COLUMN+CELL
 Lucy                              column=info:HR, timestamp=1547184039130, value=94
 Lucy                              column=info:Sales, timestamp=1547184039130, value=89
 Michael                           column=info:DB, timestamp=1547184039130, value=80
 Shelley                           column=info:Python, timestamp=1547184039130, value=80
 Will                              column=info:Perl, timestamp=1547184039130, value=85
4 row(s) in 0.0170 seconds
```

图8.12　emp_map表数据信息

从图 8.12 中可以看出，Hive 表中的 MAP 字段的值{"DB":80}转换成了 HBase 表中的 column=info:DB，value=80。

注意

这种整合方式要求 Hive 表中 MAP 类型的 Key 必须是 string 类型，否则插入数据会失败，因为 Key 将被映射为 HBase 列的名字。

8.2.3　将零售商店顾客购买统计信息存入 HBase 表

因为 HBase 表读写速度快，可以实时地响应用户的即时请求，所以在实际应用中，比较常见的一种场景就是将 Hive 分析统计的结果映射为 HBase 表。本节主要介绍如何将零售商店的数据信息分析统计后映射为 HBase 表，并供前端业务人员使用。

1. 需求

统计每位顾客的总消费额及购买商品个数，要求字段如表 8-1 所示。

表 8-1　字段要求

统计字段	字段描述	备注
name	顾客姓名	格式 "fname lname"，类型 string
total_quantity	购买商品个数	类型 string
sum_price	总消费额	类型 string

2. 实现代码

（1）首先在 Hive 中建立存储统计信息的表 shopping_statistics，然后使用 Storage Handler 将其映射为 HBase 表，列族为"statistics"，语句如下。

```
create table shopping_statistics(name string,total_quantity string,sum_price string)
stored by 'org.apache.hadoop.hive.hbase.HBaseStorageHandler'
with serdeproperties (
"hbase.columns.mapping" = ":key,statistics:totalQuantity,statistics:sumPrice"
)
tblproperties (
"hbase.table.name" = "shopping_info",
"hbase.mapred.output.outputtable" = "shopping_info"
);
```

（2）向表中插入统计所得每个顾客的信息，语句如下。

```
insert overwrite table shopping_statistics
```

```
select concat(c.customer_fname,' ',c.customer_lname) as name,
count(*) over(partition by c.customer_id) as total_quantity,
sum(oi.order_item_subtotal) over(partition by c.customer_id) as sum_price
from customers c
left join orders o on c.customer_id = o.order_customer_id
left join order_items oi on o.order_id = oi.order_item_order_id;
```

3. 运行结果

（1）Hive 中 shopping_statistics 表数据信息如图 8.13 所示。

```
+----------------------------+------------------------------------+-----------------------------+
| shopping_statistics.name   | shopping_statistics.total_quantity | shopping_statistics.sum_price |
+----------------------------+------------------------------------+-----------------------------+
| Aaron Berger               | 12                                 | 2609.63                     |
| Aaron Boyle                | 14                                 | 1779.88                     |
| Aaron Bush                 | 17                                 | 2359.66                     |
| Aaron Calhoun              | 5                                  | 949.9100000000001           |
| Aaron Carr                 | 29                                 | 5409.23                     |
| Aaron Chambers             | 1                                  | NULL                        |
| Aaron Cline                | 13                                 | 2763.7799999999997          |
| Aaron Cooke                | 10                                 | 2149.83                     |
| Aaron Cordova              | 5                                  | 999.94                      |
| Aaron Davis                | 10                                 | 1959.72                     |
+----------------------------+------------------------------------+-----------------------------+
```

图8.13 shopping_statistics表数据信息

（2）HBase 数据库中 shopping_info 表数据信息如图 8.14 所示。

```
ROW                    COLUMN+CELL
 Aaron Berger          column=statistics:sumPrice, timestamp=1547188998052, value=2609.63
 Aaron Berger          column=statistics:totalQuantity, timestamp=1547188998052, value=12
 Aaron Boyle           column=statistics:sumPrice, timestamp=1547188997598, value=1779.88
 Aaron Boyle           column=statistics:totalQuantity, timestamp=1547188997598, value=14
 Aaron Bush            column=statistics:sumPrice, timestamp=1547189001437, value=2359.66
 Aaron Bush            column=statistics:totalQuantity, timestamp=1547189001437, value=17
 Aaron Calhoun         column=statistics:sumPrice, timestamp=1547189000906, value=949.9100000000001
 Aaron Calhoun         column=statistics:totalQuantity, timestamp=1547189000906, value=5
 Aaron Carr            column=statistics:sumPrice, timestamp=1547188996478, value=5409.23
 Aaron Carr            column=statistics:totalQuantity, timestamp=1547188996478, value=29
5 row(s) in 0.0080 seconds
```

图8.14 shopping_info表数据信息

8.2.4 技能实训

统计零售商店每个顾客每天购买的商品个数及总消费额，并将统计结果映射为 HBase 数据库中的表。字段要求如表 8-2 所示。

表 8-2 每日数据统计字段要求

统计字段	字段描述	备注
name	顾客姓名	格式“fname lname”，类型 string
daily_total_quantity	每日购买商品个数	类型 string
daily_sum_price	每日总消费额	类型 string

关键步骤：

（1）使用 CREATE TABLE 命令以及 HBaseStorageHandler 创建表；

（2）设计查询语句，统计零售商店每个顾客每天购买的商品个数及总消费额；

（3）将统计数据插入到创建的表中；

（4）分别在 Hive 和 HBase 中查看表中的数据。

任务 3 使用 Phoenix 操作 HBase 数据库

【任务描述】

了解 Phoenix 工具，搭建 Phoenix 环境，并学习使用 Phoenix 操作 HBase 数据库的方法。

【关键步骤】

（1）了解 Phoenix 工具。

（2）搭建 Phoenix 环境。

（3）使用 Phoenix 操作 HBase 数据库。

8.3.1 Phoenix 简介

1. Phoenix 简介

Phoenix 最早是 saleforce 的一个开源项目，后来成为了 Apache 基金的顶级项目。它是构建在 HBase 上的一个 SQL 层，方便用户使用标准的 JDBC API 对 HBase 执行查询表、创建表、插入数据等操作。

Phoenix 可在 Hadoop 中实现联机事务处理（OLTP）和运营分析，其低延迟应用则主要通过结合以下两个优势实现。

- 具有完成 ACID 事务、标准 SQL 以及 JDBC API 的强大功能。
- 利用 HBase 进行后台存储，为 NoSQL 世界提供了迟绑定（late-bound）、读模式（schema-on-read）等灵活的功能。

目前，使用 Phoenix 的部分公司的商标如图 8.15 所示。

图8.15 使用Phoenix的部分公司的商标

2. Phoenix 架构

Phoenix 完全使用 Java 编写，接收到 SQL 查询后，会将其编译成一系列的 HBase

Scans，并会协调这些 Scans 运行，以生成常规的 JDBC 结果集。表元数据存储在 HBase 的一个表中，并对其进行了版本控制。Phoenix 架构如图 8.16 所示。

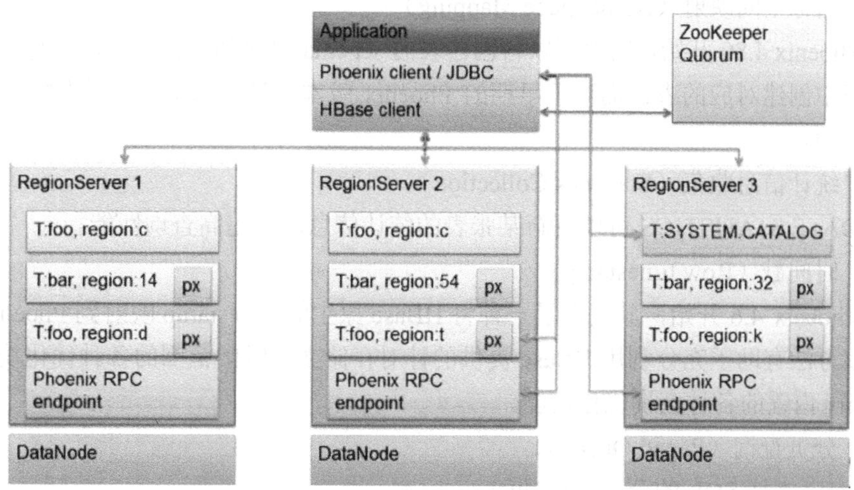

图8.16　Phoenix架构

3. Phoenix 特性

（1）事务（Transactions）

该特性还处于 beta 版，并非正式版。通过集成 Tephra，Phoenix 可以支持 ACID 特性。Tephra 也是 Apache 的一个项目，瞄准事务管理，它可以在像 HBase 这样的分布式数据存储平台上提供全局一致事务。HBase 本身在行层次和区层次上支持强一致性，Tephra 又额外提供交叉区、交叉表的一致性来支持可扩展性。

（2）用户定义函数（User-Defined Functions，UDFs）

Phoenix 从 4.4.0 版本开始支持用户自定义函数。用户可以创建临时或永久的用户自定义函数，且这些函数可以像内置的 SELECT、CREATE、UPSERT、DELETE 一样被调用。临时函数仅针对特定的会话或连接，对其他会话或连接则不可见。永久函数的元信息会被存储在一张叫作 SYSTEM.FUNCTION 的系统表中，对任何会话或连接均可见。

（3）二级索引（Secondary Indexing）

在 HBase 中，有一个单一的按照字典序排序的 Row Key 索引，当使用 Row Key 进行数据查询时查询速度较快，但是如果不使用 Row Key 进行数据查询，则会使用 Filter 对全表进行扫描，这在很大程度上会降低检索性能。Phoenix 提供了二级索引技术以应对不使用 Row Key 时进行检索的场景。Phoenix 支持的二级索引技术详情可扫描二维码获取。

Phoenix 二级索引技术详情

（4）存储格式（Storage Format）

Phoenix 4.10 通过以下两个方式减少了磁盘存储大小，提高了性能。

➢ 在 Phoenix 列名和相应的 HBase 列限定符之间引入了一个间接层。

➢ 开始支持面向不可变表的新编码方案，该方案会将所有值打包到每个列族的单元格中。

（5）原子插入（Atomic Upsert）

此特性可以启用 HBase 在增量、检查和插入数据时的原子插入功能。

（6）名称空间映射（NameSpace Mapping）

从 Phoenix 4.80 开始，用户可以将创建表的 schema 映射到名称空间，以便在相应的命名空间中创建对应的表；而对于早期的 Phoenix 版本来说，每个表都是在默认名称空间中创建的。

（7）统计信息收集（Statistics Collection）

UPDATE STATISTICS 可以更新某张表的统计信息，以提高查询性能。

（8）时间戳（Row timestamp）

从 Phoenix 4.6 开始，其提供了一种将 HBase 原生的 timestamp 映射到 Phoenix 列的方法，该方法有助于充分利用 HBase 提供的针对存储文件时间范围的各种优化功能，以及 Phoenix 内置的各种查询优化功能。

（9）分页查询（Paged Queries）

Phoenix 支持 SQL 的分页查询功能。

（10）散步表（Salted Tables）

如果 Row Key 是自动增长的，那么 HBase 的顺序写就会引发数据热点问题，而 Phoenix 的 Salted Tables 技术可以解决 Region Server 的数据热点问题。

（11）跳跃扫描（Skip Scan）

可以在范围扫描的时候提高性能。

8.3.2 搭建 Phoenix CDH 环境

1. 安装包下载

本书采用的是 CDH 版本的 apache-phoenix-4.14.0-cdh5.14.2，读者可以自行在官网下载其安装包。安装环境要求如下。

（1）操作系统：CentOS 7.5。

（2）Java 环境：Java 8。

（3）Hadoop 环境：hadoop-2.6.0-cdh5.14.2。

（4）HBase 环境：hbase-1.2.0-cdh5.14.2。

2. 解压安装

（1）下载并解压安装包

```
$ tar -zxf apache-phoenix-4.14.0-cdh5.14.2-bin.tar
```

（2）将解压后的文件复制到“/opt”目录下

```
$ sudo mv apache-phoenix-4.14.0-cdh5.14.2-bin/ /opt/apache-phoenix-4.14.0-cdh5.14.2-bin
```

如果采用的是普通用户安装模式，则需要加 sudo 才能操作“/opt”目录。

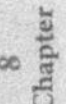

3. Phoenix CDH 环境搭建

Phoenix CDH 环境搭建的前提是 Hadoop 集群、ZooKeeper、HBase 均安装成功。

（1）将 Phoenix 的安装目录添加到环境变量当中，命令如下。

```
$ vi ~/.bashrc
export PHOENIX_HOME= /opt/apache-phoenix-4.14.0-cdh5.14.2-bin
export PHOENIX_CLASSPATH=$PHOENIX_HOME
export PATH= $PATH:$HIVE_HOME/bin
```

添加完成以后，执行如下命令，使环境变量生效。

```
$ source ~/.bashrc
```

（2）将${PHOENIX_HOME}/phoenix-4.14.0-cdh5.14.2-server.jar 拷贝到 HBase 节点的 ${HBASE_HOME}/lib 目录下。

```
$ cp ${PHOENIX_HOME}/phoenix-4.14.0-cdh5.14.2-server.jar ${HBASE_HOME}/lib
```

（3）重启 HBase 集群。

```
$ ${HBASE_HOME}/bin/stop-hbase.sh
$ ${HBASE_HOME}/bin/start-hbase.sh
```

（4）启动 Phoenix，验证其安装成功。

```
$ cd ${PHOENIX_HOME}/bin
$ ./sqlline.py
```

启动结果如图 8.17 所示。

```
Setting property: [incremental, false]
Setting property: [isolation, TRANSACTION_READ_COMMITTED]
issuing: !connect jdbc:phoenix: none none org.apache.phoenix.jdbc.Ph
enixDriver
Connecting to jdbc:phoenix:
SLF4J: Class path contains multiple SLF4J bindings.
SLF4J: Found binding in [jar:file:/opt/apache-phoenix-4.14.0-cdh5.14
2-bin/phoenix-4.14.0-cdh5.14.2-client.jar!/org/slf4j/impl/StaticLogg
rBinder.class]
SLF4J: Found binding in [jar:file:/opt/hadoop-2.6.0-cdh5.14.2/share/
adoop/common/lib/slf4j-log4j12-1.7.5.jar!/org/slf4j/impl/StaticLogge
Binder.class]
SLF4J: See http://www.slf4j.org/codes.html#multiple_bindings for an
xplanation.
SLF4J: Actual binding is of type [org.slf4j.impl.Log4jLoggerFactory]
19/01/18 16:58:07 WARN util.NativeCodeLoader: Unable to load native-
adoop library for your platform... using builtin-java classes where
pplicable
Connected to: Phoenix (version 4.14)
Driver: PhoenixEmbeddedDriver (version 4.14)
Autocommit status: true
Transaction isolation: TRANSACTION_READ_COMMITTED
Building list of tables and columns for tab-completion (set fastconn
ct to true to skip)...
163/163 (100%) Done
Done
sqlline version 1.2.0
0: jdbc:phoenix:>
```

图8.17　Phoenix启动结果

至此，Phoenix CDH 安装已完成。

4. Phoenix 操作 HBase 表

Phoenix 安装完成以后，即可使用 Phoenix 来操作 HBase 表。

（1）列出所有表

示例 8-7

使用 Phoenix 列出 HBase 数据库中的所有表。

关键代码：

```
!tables
```

代码运行结果如图 8.18 所示。

TABLE_CAT	TABLE_SCHEM	TABLE_NAME	TABLE_TYPE	REMARKS	TYPE_NAME	SELF_REFERENCING_COL_NAME	REF_GENERATION	INDEX_STA
	SYSTEM	CATALOG	SYSTEM TABLE					
	SYSTEM	FUNCTION	SYSTEM TABLE					
	SYSTEM	LOG	SYSTEM TABLE					
	SYSTEM	SEQUENCE	SYSTEM TABLE					
	SYSTEM	STATS	SYSTEM TABLE					
		dv	VIEW					
		region_st	VIEW					
		test	VIEW					
		user_active	VIEW					
		user_olst	VIEW					
		useradd_al	VIEW					

图8.18　使用Phoenix列出所有表代码运行结果

（2）创建表

示例 8-8

使用 Phoenix 命令行在 HBase 数据库中创建 users 表。

关键代码：

```
create table "users" ("id" integer not null primary key, "cf"."name" varchar, "cf"."age" integer);
```

（3）查看表结构

示例 8-9

使用 Phoenix 命令行查看 users 表的结构信息。

关键代码：

```
!describe "users"
```

代码运行结果如图 8.19 所示。

TABLE_CAT	TABLE_SCHEM	TABLE_NAME	COLUMN_NAME	DATA_TYPE	TYPE_NAME	COLUMN_SIZE	BUFFER_LENGTH	DECIMAL_DIGITS	NUM_PR
		users	id	4	INTEGER	null	null	null	null
		users	name	12	VARCHAR	null	null	null	null
		users	age	4	INTEGER	null	null	null	null

图8.19　使用Phoenix查看表结构代码运行结果

（4）插入数据

示例 8-10

使用 Phoenix 命令行在 users 表中插入数据。

关键代码：

```
upsert into "users" values(1,'zhangsan',27);
```

（5）查看表数据

示例 8-11

使用 Phoenix 命令行查看 users 表数据。

关键代码：

```
select * from "users";
```

代码运行结果如图 8.20 所示。

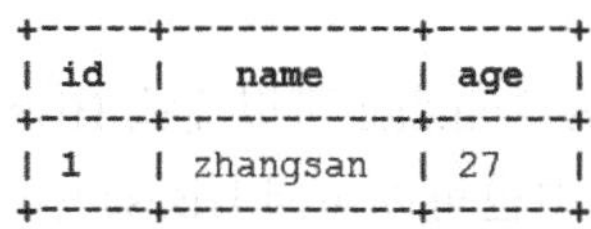

id	name	age
1	zhangsan	27

图8.20　使用Phoenix查看表数据代码运行结果

（6）删除表

示例 8-12

使用 Phoenix 命令行删除 users 表。

关键代码：

```
drop table "users";
!tables
```

代码运行结果如图 8.21 所示。

TABLE_CAT	TABLE_SCHEM	TABLE_NAME	TABLE_TYPE	REMARKS	TYPE_NAME	SELF_REFERENCING_COL_NAME	REF_GENERATION	INDEX_STA
	SYSTEM	CATALOG	SYSTEM TABLE					
	SYSTEM	FUNCTION	SYSTEM TABLE					
	SYSTEM	LOG	SYSTEM TABLE					
	SYSTEM	SEQUENCE	SYSTEM TABLE					
	SYSTEM	STATS	SYSTEM TABLE					
		person	TABLE					
		dv	VIEW					
		region_st	VIEW					
		test	VIEW					
		user_active	VIEW					
		user_olst	VIEW					
		useradd_al	VIEW					

图8.21　使用Phoenix删除表代码运行结果

从图 8.21 中可以看出，users 表已经被删除。

更多 Phoenix 操作介绍请扫描二维码获取。

8.3.3　技能实训

Phoenix 操作

使用 Phoenix 命令行方式完成 HBase 表管理操作。

➢　在 HBase 中创建学生表 students。

➢　完成对学生表的管理操作（即查看表结构在表中插入数据、查看表数据、删除表）。

关键步骤：

（1）使用 CREATE TABLE 命令创建表；

（2）使用!DESCRIBE 命令查看表结构信息；

（3）使用 UPSERT 命令在表中插入数据；

（4）使用 SELECT 命令查看表数据；

（5）使用 DROP TABLE 命令删除表。

本章小结

➢ Hive 提供了 StorageHandler 接口，用户可以通过实现该接口中的方法编写定制化的存储处理器。

➢ Hive 提供的 HBaseStorageHandler 可以完成 Hive 与 HBase 的整合。在 HBase 中可以访问 Hive 中的表，在 Hive 中也可以访问 HBase 中的表。

➢ 除了 Hive 可以为 HBase 提供类 SQL 查询外，Phoenix 也可以作为 HBase 的 SQL 层，通过使用标准 SQL 语句访问 HBase 数据库。

本章作业

一、简答题

1. 简述 Hive 与 HBase 集成后的应用场景。

2. 简述 HiveStorageHandler 原理。

二、编码题

1. 基于 retail_db 数据库中的数据表（products、orders、order_items），在 Hive 中建立统计分析表。

要求：

➢ 统计每个商品的销售总数。

➢ 用户可以通过 HBase 访问到该统计表。

统计表字段说明如表 8-3 所示。

表 8-3　商品销售总数统计表字段说明

统计字段	字段描述	备注
product_name	商品名称	类型 string
total_quantity	商品销售总数	类型 string

2. 使用 Phoenix 建立 HBase 表 orders，并完成该表的管理操作。表字段说明参考 Hive 数据仓库中“retail_db”数据库中的表结构信息。

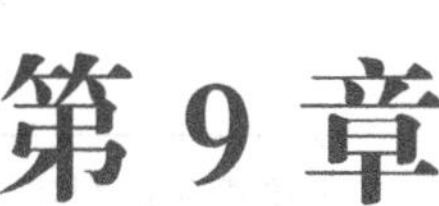

第 9 章 数据迁移框架 Sqoop

技能目标

- 理解 Sqoop 的原理。
- 掌握 Sqoop 的结构。
- 熟练使用 Sqoop 进行数据导入导出。
- 掌握 Sqoop Job 的配置。

本章任务

任务 1　使用 Sqoop 完成 Hadoop 与 MySQL 间的数据迁移。

任务 2　使用 Sqoop Job 完成 Hive 与 MySQL 间的数据迁移。

本章资源下载

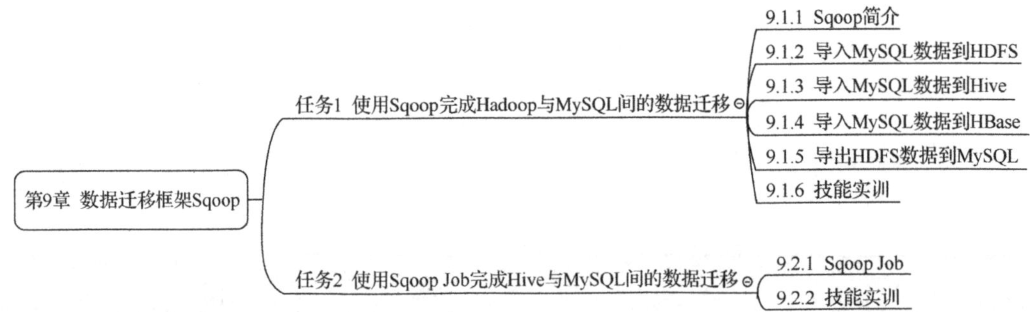

Sqoop 是大数据环境下企业中经常使用的数据迁移工具之一，其可以非常便捷地实现数据在大数据环境下的导入/导出。本章将介绍如何使用 Sqoop 实现数据迁移。

任务 1 使用 Sqoop 完成 Hadoop 与 MySQL 间的数据迁移

【任务描述】

了解 Sqoop 的产生背景及其环境搭建方法，掌握如何使用 Sqoop 在 Hadoop 与关系型数据库之间传输数据。

【关键步骤】

（1）导入 MySQL 数据到 HDFS、Hive、HBase。

（2）导出 HDFS 数据并将其输入到 MySQL。

9.1.1 Sqoop 简介

1．Sqoop 产生背景

在工作中，经常会遇到如下数据迁移的场景。

场景一：将关系型数据库的某张表中的数据抽取到 Hadoop（HDFS/Hive/HBase）上。

场景二：将 Hadoop 上的数据导出到关系型数据库中。

那么如何解决上述两类问题呢？通常情况下会通过开发 MapReduce 应用来实现。

导入：MapReduce 输入为 DBInputFormat 类型，输出为 TextOutputFormat。

导出：MapReduce 输入为 TextInputFormat 类型，输出为 DBOutputFormat。

在上述两个场景中使用 MapReduce 应用时会存在如下问题：每次应用都需要编写 MapReduce 程序，效率很低。在没有出现 Sqoop 之前，实际生产过程中有很多类似的需求须通过编写 MapReduce 去满足。在长期的 MapReduce 编写过程中形成了一个工具，随后将该工具代码整理成一个框架并逐渐完善，最终才有了 Sqoop 的诞生。

2．Sqoop 概念

Sqoop 由 SQL 与 Hadoop 组合而成，是基于 Hadoop 的数据传输和迁移工具。其目前是 Apache 顶级项目，主要用于在关系型数据库、Hadoop、数据仓库、NoSQL 系统中传递数据。通过 Sqoop 我们可以方便地将数据从关系型数据库中导入到 HDFS、Hive 、HBase，或者将数据从 HDFS 中导出到关系型数据库。使用 Sqoop 导入/导出处理流程如

图 9.1 所示。

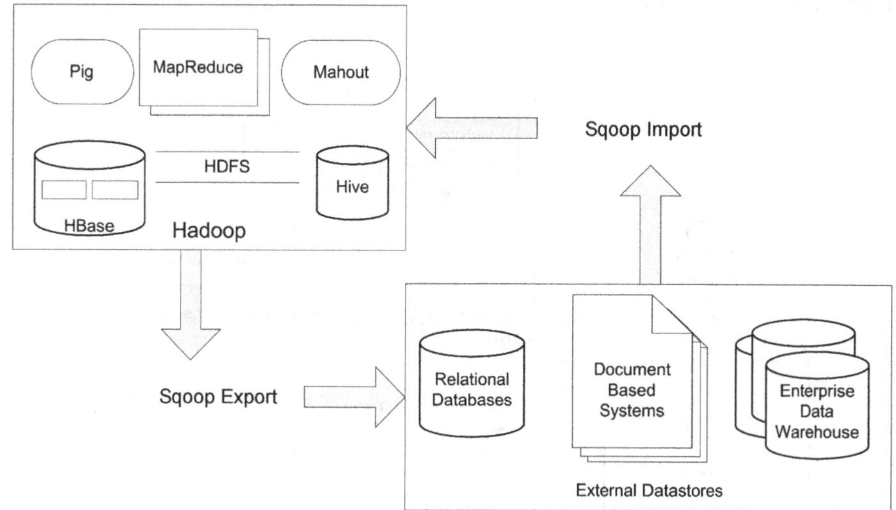

图9.1　Sqoop导入/导出处理流程

Sqoop 在 HDFS 生态圈中的位置如图 1.1 所示。

Sqoop 是连接传统关系型数据库和 Hadoop 的桥梁，其无须开发人员编写相应的 MapReduce 代码，而只须简单地编写并配置脚本即可大大提升开发效率。

3．Sqoop 版本介绍

2012 年 3 月，Sqoop 从 Apache 的孵化器中“毕业”，成为 Apache 的顶级项目。Sqoop 版本发展历程如图 9.2 所示。

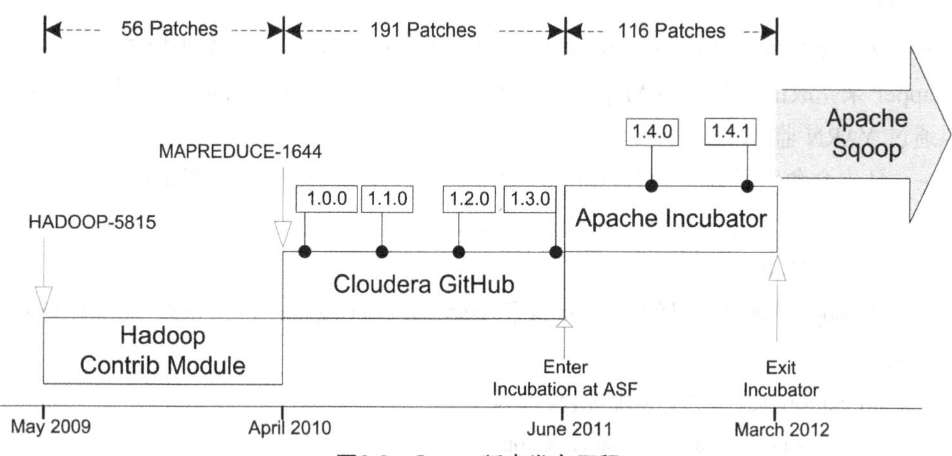

图9.2　Sqoop版本发布历程

Sqoop 的版本到目前为止，主要分为 Sqoop1 和 Sqoop2，Sqoop1.4.7 之前的所有版本称为 Sqoop1，之后的 Sqoop1.991、Sqoop1.99.2、Sqoop1.99.3 版本称为 Sqoop2。目前 Sqoop1 的稳定版本是 Sqoop1.4.6，Sqoop2 的最新版本是 Sqoop 1.99.7。两个版本在架构和使用上有很大的区别，本书中我们选择 Sqoop1.x 进行相关内容的讲解。

4. Sqoop 架构

Sqoop1.x 架构如图 9.3 所示。

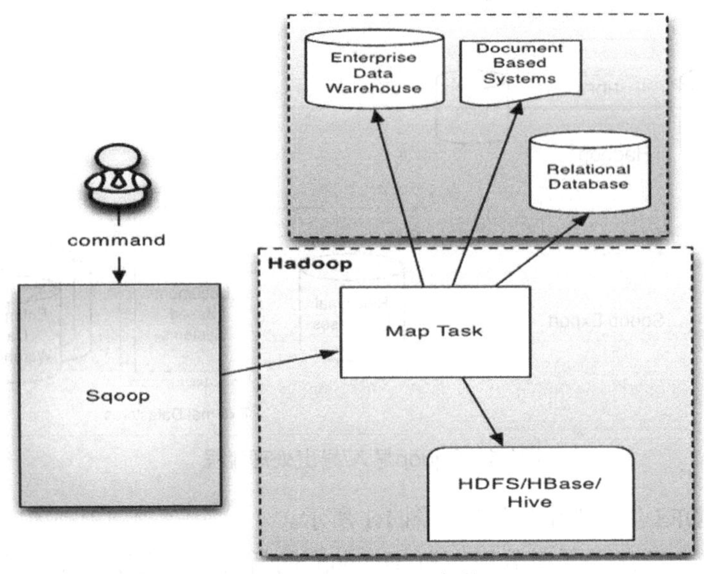

图9.3 Sqoop1.x架构

Sqoop 的架构非常简单，其整合了 Hive、HBase 等，通过 Map 任务来传输数据。Map 负责数据的加载与转换，以及将转换以后的数据存储到 HDFS、HBase 或者 Hive 中。

➢ 从工作模式角度看，Sqoop 是基于客户端模式的，用户在使用 Sqoop 时，只需要在一台机器上运行即可。

➢ 从 MapReduce 角度看，Sqoop 只提交一个 Map 作业，数据的传输和转换都是使用 Mapper 来完成的，而且该 Map 作业仅有 Mapper 而不需要 Reducer，在执行 Sqoop 时可以通过 YARN 监控页面查看。

➢ 从安全角度看，在执行 Sqoop 时须将用户名或者密码显示指定，也可以将其配置在配置文件中，总之 Sqoop 安全性并不高。

5. Sqoop 部署

安装 Sqoop 的前提是已经具备 Java 和 Hadoop 的环境。在官网下载 Sqoop 安装包。安装步骤如下。

（1）将 Sqoop 安装包解压到安装目录。

```
$ tar -zxvf sqoop-1.4.6-cdh5.14.2.tar.gz -C /opt/
```

（2）添加环境变量。

```
$ vi ~/.bash_profile
export SQOOP_HOME=/opt/sqoop-1.4.6-cdh5.14.2
export PATH=$SQOOP_HOME/bin:$PATH
```

（3）修改 Sqoop 配置文件：$SQOOP_HOME/conf/sqoop-env.sh。

```
$ cd $SQOOP_HOME/conf
```

```
$ cp sqoop-env-template.sh sqoop-env.sh
$ vi sqoop-env.sh
#指定到 Hadoop 安装路径
export HADOOP_COMMON_HOME=/opt/hadoop-2.6.0-cdh5.14.2
export HADOOP_MAPRED_HOME=/opt/hadoop-2.6.0-cdh5.14.2
#指定到 Hive 安装路径
export HIVE_HOME=/opt/hive-1.1.0-cdh5.14.2
```

（4）拷贝 MySQL 驱动 JAR 文件到$SQOOP_HOME/lib 下。

```
$ cp ~/software/mysql-connector-java-5.1.44-bin.jar $SQOOP_HOME/lib/
```

（5）Sqoop 配置验证。

```
$ sqoop version
```

6. Sqoop 简单使用

示例 9-1

Sqoop 帮助命令的使用。

关键代码：

```
$ sqoop help
```

代码运行结果如图 9.4 所示。

```
usage: sqoop COMMAND [ARGS]

Available commands:
  codegen            Generate code to interact with database records
  create-hive-table  Import a table definition into Hive
  eval               Evaluate a SQL statement and display the results
  export             Export an HDFS directory to a database table
  help               List available commands
  import             Import a table from a database to HDFS
  import-all-tables  Import tables from a database to HDFS
  import-mainframe   Import datasets from a mainframe server to HDFS
  job                Work with saved jobs
  list-databases     List available databases on a server
  list-tables        List available tables in a database
  merge              Merge results of incremental imports
  metastore          Run a standalone Sqoop metastore
  version            Display version information
```

图9.4　sqoop help运行结果

示例 9-2

查看 Sqoop 版本。

关键代码：

```
$ sqoop version
```

代码运行结果如图 9.5 所示。

```
19/01/10 11:31:19 INFO sqoop.Sqoop: Running Sqoop version: 1.4.6-cdh5.14.2
Sqoop 1.4.6-cdh5.14.2
git commit id
Compiled by jenkins on Tue Mar 27 13:19:49 PDT 2018
```

图9.5　sqoop version运行结果

示例 9-3

使用 Sqoop 获取指定 URL 的数据库。

分析：

根据 sqoop help 命令可知应使用 list-databases 命令来完成该功能，并确定需要哪些参数。

关键代码：

```
$ sqoop help list-databases
$ sqoop list-databases \
--connect jdbc:mysql://localhost:3306 \
--username root \
--password 123456
```

运行上述代码将输出 MySQL 中的所有数据库名称。

示例 9-4

使用 Sqoop 获取指定 URL 的数据库中的所有表。

关键代码：

```
$ sqoop list-tables \
--connect jdbc:mysql://localhost:3306/test \
--username root \
--password 123456
```

运行上述代码将输出“test”数据库中所有表名称。

Sqoop 提供了多种工具以完成数据迁移，常用的也是本书主要介绍的工具包括 sqoop-import、sqoop-export、Sqoop Job，其他工具介绍请扫描二维码获取。

Sqoop 其他工具介绍

9.1.2 导入 MySQL 数据到 HDFS

1. MySQL 数据准备

（1）创建 Sqoop 数据库。

```
CREATE DATABASE sqoop;
```

（2）创建部门表 dept 并向其中导入数据。

```
DROP TABLE dept;
CREATE TABLE dept(DEPTNO int(2) PRIMARY KEY,DNAME VARCHAR(14) ,LOC VARCHAR(13) ) ;
INSERT INTO dept VALUES (10,'ACCOUNTING','NEW YORK');
INSERT INTO dept VALUES (20,'RESEARCH','DALLAS');
INSERT INTO dept VALUES (30,'SALES','CHICAGO');
INSERT INTO dept VALUES (40,'OPERATIONS','BOSTON');
```

（3）创建员工表 emp 并向其中导入数据。

```
DROP TABLE emp;
CREATE TABLE emp(
EMPNO int(4) PRIMARY KEY,
ENAME VARCHAR(10),
```

```
JOB VARCHAR(9),
MGR int(4),
HIREDATE DATE,
SAL int(7),
COMM int(7),
DEPTNO int(2),
FOREIGN KEY(DEPTNO) REFERENCES DEPT(DEPTNO));
//导入数据
INSERT INTO emp VALUES(7369,'SMITH','CLERK',7902,'1980-12-17',800,NULL,20);
INSERT INTO emp VALUES(7499,'ALLEN','SALESMAN',7698,'1981-2-20',1600,300,30);
INSERT INTO emp VALUES(7521,'WARD','SALESMAN',7698,'1981-2-22',1250,500,30);
INSERT INTO emp VALUES(7566,'JONES','MANAGER',7839,'1981-4-2',2975,NULL,20);
INSERT INTO emp VALUES(7654,'MARTIN','SALESMAN',7698,'1981-9-28',1250,1400,30);
INSERT INTO emp VALUES(7698,'BLAKE','MANAGER',7839,'1981-5-1',2850,NULL,30);
INSERT INTO emp VALUES(7782,'CLARK','MANAGER',7839,'1981-6-9',2450,NULL,10);
INSERT INTO emp VALUES(7788,'SCOTT','ANALYST',7566,'87-7-13',3000,NULL,20);
INSERT INTO emp VALUES(7839,'KING','PRESIDENT',NULL,'1981-11-17',5000,NULL,10);
INSERT INTO emp VALUES(7844,'TURNER','SALESMAN',7698,'1981-9-8',1500,0,30);
INSERT INTO emp VALUES(7876,'ADAMS','CLERK',7788,'87-7-13',1100,NULL,20);
INSERT INTO emp VALUES(7900,'JAMES','CLERK',7698,'1981-12-3',950,NULL,30);
INSERT INTO emp VALUES(7902,'FORD','ANALYST',7566,'1981-12-3',3000,NULL,20);
INSERT INTO emp VALUES(7934,'MILLER','CLERK',7782,'1982-1-23',1300,NULL,10);
```

2. sqoop-import

import 工具提供（从 RDBMS 到 HDFS）导入单个表的功能，表的每一行都在 HDFS 中对应一条单独的记录，记录可以存储为文本、Avro 或 SequenceFile 格式。

【语法】

```
$ sqoop import [generic-args] [import-args]
```

sqoop-import 参数包括通用参数（如表 9-1 所示）、导入控制参数（如表 9-2 所示）、输出格式控制参数（如表 9-3 所示）、输入格式控制参数（如表 9-4 所示）、Hive 参数（如表 9-5 所示）、HBase 参数（如表 9-6 所示）等。

表 9-1　通用参数

参数	描述
--connect <jdbc-uri>	指定 JDBC 连接串
--username <username>	用户名
--password <password>	密码
--password-file	包含密码的文件

表 9-2　导入控制参数

参数	描述
--append	追加数据至已存在的 HDFS 数据集
--as-avrodatafile	导入数据至 Avro 文件（默认是文本文件）

续表

参数	描述
--as-parquetfile	导入数据至 Parquet 文件
--as-sequencefile	导入数据至 Sequence 文件
--columns <col,col,col...>	指定导入的列
--compression-codec <codec>	压缩编码
--delete-target-dir	如果导入目录文件夹存在则删除
--direct	如果数据库存在则使用直接连接
-e,--query <statement>	如果是 SQL 语句则查询结果将被导入
-m,--num-mappers <n>	导入时的并行 Map 任务的数量
--mapreduce-job-name <name>	作业名称
--table <table-name>	读取的表名
--target-dir <dir>	将要导入的 HDFS 的目录
--where <where clause>	条件过滤
-z,--compress	启用压缩

表 9-3　输出格式控制参数

参数	描述
--fields-terminated-by <char>	设置字段分隔符
--lines-terminated-by <char>	设置行分隔符

表 9-4　输入格式控制参数

参数	描述
--input-enclosed-by <char>	设置输入字符包围符
--input-escaped-by <char>	设置输入转义符
--input-fields-terminated-by <char>	设置输入字段分隔符
--input-lines-terminated-by <char>	设置输入行分隔符

表 9-5　Hive 参数

参数	描述
--create-hive-table	自动创建 Hive 表
--hive-database <database-name>	设置 Hive 数据库名
--hive-import	导入 RDBMS 表至 Hive 中
--hive-overwrite	如果数据在 Hive 表中存在则覆盖
--hive-partition-key <partition-key>	分区键
--hive-partition-value <partition-value>	分区值
--hive-table <table-name>	指定导入 Hive 的表

表 9-6　HBase 参数

参数	描述
--column-family <family>	设置导入的目标列族
--hbase-create-table	自动创建 HBase 表
--hbase-row-key <col>	指定哪一列作为 RowKey
--hbase-table	指定导入的 HBase 表名
--hbase-bulkload	启用批量加载

示例 9-5

导入 MySQL 中 sqoop.emp 表的所有字段至 HDFS。

关键代码：

```
$ sqoop import \
--connect jdbc:mysql://master:3306/sqoop \
--username root --password 123456 \
--table emp
-m 1
```

注意

- 须使用--connect 来指定要导入数据的数据库。
- 须使用--username 和--password 来指定数据库的用户名和密码。
- 须使用--table 来指定须被导入的数据表。
- 须使用-m 来指定导入数据的并行度，Sqoop 默认的并行度是 4，此处我们将其设置为 1；在 HDFS 上最终输出的文件个数就是并行度的个数。
- 使用 Sqoop 从关系型数据库中导入数据到 HDFS 时，默认的路径是/user/用户名/表名。
- Sqoop 默认从关系型数据库中导入数据到 HDFS 的分隔符是逗号。

执行完上述代码之后可以在 HDFS 上查看导出的结果。

```
// 查看在 HDFS 上的结果导出路径
$ hdfs dfs -ls /user/root/emp/
// 查看导出结果
$ hdfs dfs -text /user/root/emp/part-*
```

导出结果如图 9.6 所示。

当对示例 9-5 再次执行 Sqoop 导入操作时，会有如下错误信息输出：输出文件已经存在。对于这个错误输出，有 MapReduce 编程基础的读者应该已经知道了其出现的原因：当 MapReduce 作业执行输出时，如果输出目录已经存在，那么就会报该错。解决办法是可以手工将存在的路径删除，但是每次都进行手工删除是非常麻烦的，为此 Sqoop 提供了可以通过参数（--delete-target-dir）指定的方式使输出路径自动删除这一功能。指定了该参数，同一 Sqoop 脚本就可以执行多次而不会再报错，代码如下所示。

```
$ sqoop import \
--connect jdbc:mysql://master:3306/sqoop \
--username root --password 123456 \
--table emp
-m 1
--delete-target-dir
```

```
[root@master ~]# hadoop fs -cat /user/root/emp/*
19/01/10 14:40:02 WARN util.NativeCodeLoader: Unabl
a classes where applicable
7369,SMITH,CLERK,7902,1980-12-17,800,null,20
7499,ALLEN,SALESMAN,7698,1981-02-20,1600,300,30
7521,WARD,SALESMAN,7698,1981-02-22,1250,500,30
7566,JONES,MANAGER,7839,1981-04-02,2975,null,20
7654,MARTIN,SALESMAN,7698,1981-09-28,1250,1400,30
7698,BLAKE,MANAGER,7839,1981-05-01,2850,null,30
7782,CLARK,MANAGER,7839,1981-06-09,2450,null,10
7788,SCOTT,ANALYST,7566,1987-07-13,3000,null,20
7839,KING,PRESIDENT,null,1981-11-17,5000,null,10
7844,TURNER,SALESMAN,7698,1981-09-08,1500,0,30
7876,ADAMS,CLERK,7788,1987-07-13,1100,null,20
7900,JAMES,CLERK,7698,1981-12-03,950,null,30
7902,FORD,ANALYST,7566,1981-12-03,3000,null,20
7934,MILLER,CLERK,7782,1982-01-23,1300,null,10
```

图9.6 sqoop将emp表导入至HDFS

另外，对于 MySQL 连接字符串也需要注意，应该为其指定实际的 IP 地址或者主机名称。因为当 Sqoop 开始执行 MapReduce 时，具体的任务需要从 MySQL 服务器中获取数据，所以 MySQL 地址必须明确，特别注意在分布式 Hadoop 环境下其不能为"localhost"。

示例 9-6

导入指定字段的表并指定目标地址。

关键代码：

```
$ sqoop import \
--connect jdbc:mysql://master:3306/sqoop \
--username root --password 123456 \
--columns "EMPNO,ENAME,JOB,SAL,COMM" \
--target-dir    emp_column \
--mapreduce-job-name FromMySQLToHDFS \
--table emp    -m 1 \
--delete-target-dir
```

说明：

- 须使用--columns 来指定要导入的字段。
- 须使用--target-dir 来指定导入数据的 HDFS 上的目标地址。
- 须使用--mapreduce-job-name 来指定该作业的名称，可以通过 YARN 页面查看之。

导入结果如图 9.7 所示。

```
[root@master ~]# hdfs dfs -cat /user/root/emp_column/*
19/01/10 15:04:10 WARN util.NativeCodeLoader: Unable to
a classes where applicable
7369,SMITH,CLERK,800,null
7499,ALLEN,SALESMAN,1600,300
7521,WARD,SALESMAN,1250,500
7566,JONES,MANAGER,2975,null
7654,MARTIN,SALESMAN,1250,1400
7698,BLAKE,MANAGER,2850,null
7782,CLARK,MANAGER,2450,null
7788,SCOTT,ANALYST,3000,null
7839,KING,PRESIDENT,5000,null
7844,TURNER,SALESMAN,1500,0
7876,ADAMS,CLERK,1100,null
7900,JAMES,CLERK,950,null
7902,FORD,ANALYST,3000,null
7934,MILLER,CLERK,1300,null
```

图9.7　导入指定字段的表并指定目标地址

示例 9-7

导入表数据并指定压缩格式以及存储格式。

关键代码：

```
$ sqoop import \
--connect jdbc:mysql://master:3306/sqoop \
--username root --password 123456 \
--columns "EMPNO,ENAME,JOB,SAL,COMM" \
--mapreduce-job-name FromMySQLToHDFS \
--target-dir    emp_parquet \
--as-parquetfile \
--compression-codec org.apache.hadoop.io.compress.SnappyCodec \
--table emp    -m 1 \
--delete-target-dir
```

注意

➢ 须使用--as-parquet 来指定导出格式为 Parquet 格式，当然也可指定导出格式为 SequenceFile 等其他格式。

➢ 须使用--compression-codec 来指定压缩使用的 codec 编码；因为在 Sqoop 中默认是使用压缩的，所以此处只须指定 codec 编码即可。

导入结果如图 9.8 所示。

```
2019-01-10 15:08 /user/root/emp_parquet/.metadata
2019-01-10 15:09 /user/root/emp_parquet/.signals
2019-01-10 15:09 /user/root/emp_parquet/6cff5a00-8b3b-4be7-a61f-3d4979e8e217.parquet
```

图9.8　导入表数据并指定压缩格式与存储格式

示例 9-8

导入表数据并使用指定的分隔符。

关键代码：

```
$ sqoop import \
--connect jdbc:mysql://master:3306/sqoop \
--username root --password 123456 \
--columns "EMPNO,ENAME,JOB,SAL,COMM" \
--mapreduce-job-name FromMySQLToHDFS \
--target-dir    emp_column_split \
--fields-terminated-by '\t' --lines-terminated-by '\n' \
--table emp    -m 1 \
--delete-target-dir
```

注意

➢ 须使用--fields-terminated-by 来设置字段之间的分隔符。

➢ 须使用--lines-terminated-by 来设置行之间的分隔符。

导入结果如图 9.9 所示。

```
[root@master ~]# hdfs dfs -text /user/root/emp_column_split/part-*
19/01/10 15:16:59 WARN util.NativeCodeLoader: Unable to load native
a classes where applicable
7369    SMITH   CLERK   800     null
7499    ALLEN   SALESMAN        1600    300
7521    WARD    SALESMAN        1250    500
7566    JONES   MANAGER 2975    null
7654    MARTIN  SALESMAN        1250    1400
7698    BLAKE   MANAGER 2850    null
7782    CLARK   MANAGER 2450    null
7788    SCOTT   ANALYST 3000    null
7839    KING    PRESIDENT       5000    null
7844    TURNER  SALESMAN        1500    0
7876    ADAMS   CLERK   1100    null
7900    JAMES   CLERK   950     null
7902    FORD    ANALYST 3000    null
7934    MILLER  CLERK   1300    null
```

图9.9　导入表数据并使用指定的分隔符

示例 9-9

导入指定条件的数据。

关键代码：

```
$ sqoop import \
--connect jdbc:mysql://master:3306/sqoop \
--username root --password 123456 \
--columns "EMPNO,ENAME,JOB,SAL,COMM" \
--mapreduce-job-name FromMySQLToHDFS \
--target-dir    emp_column_where \
--fields-terminated-by '\t' --lines-terminated-by '\n' \
--where 'SAL>2000' \
--table emp    -m 1 \
--delete-target-dir
```

注意

须使用--where 条件来指定 emp 表中满足条件的数据。

导入结果如图 9.10 所示。

```
7566	JONES	MANAGER 2975	null
7698	BLAKE	MANAGER 2850	null
7782	CLARK	MANAGER 2450	null
7788	SCOTT	ANALYST 3000	null
7839	KING	PRESIDENT	5000	null
7902	FORD	ANALYST 3000	null
```

图9.10　导入指定条件的数据

示例 9-10

导入指定查询语句的数据。

关键代码：

```
$ sqoop import \
--connect jdbc:mysql://master:3306/sqoop \
--username root --password 123456 \
--mapreduce-job-name FromMySQLToHDFS \
--target-dir    emp_column_query \
--fields-terminated-by '\t' --lines-terminated-by '\n' \
--query 'SELECT * FROM emp WHERE empno>=7900 AND $CONDITIONS' \
-m 1 \
--delete-target-dir
```

注意

➢ 若使用--query 参数来指定查询语句，就能将 query 中的查询结果导入到 HDFS 中。

➢ 在 sqoop import 中须使用--query 来指定 SQL 的条件中需要添加$CONDITIONS，这是固定写法。

导入结果如图 9.11 所示。

```
7900	JAMES	CLERK	7698	1981-12-03	950	null	30
7902	FORD	ANALYST 7566	1981-12-03	3000	null	20
7934	MILLER	CLERK	7782	1982-01-23	1300	null	10
```

图9.11　导入指定查询语句的数据

示例 9-11

使用 eval。

分析：

eval 可以执行 SQL 语句并显示结果。

关键代码：

```
$ sqoop eval --connect jdbc:mysql://master:3306/sqoop \
```

```
--username root --password 123456 \
--query 'select * from emp where deptno=10'
```

代码运行结果如图 9.12 所示。

```
-------------------------------------------------------------------------------------------
| EMPNO | ENAME      | JOB       | MGR  | HIREDATE   | SAL     | COMM     | DEPTNO |
-------------------------------------------------------------------------------------------
| 7782  | CLARK      | MANAGER   | 7839 | 1981-06-09 | 2450    | (null)   | 10 |
| 7839  | KING       | PRESIDENT | (null) | 1981-11-17 | 5000    | (null)   | 10 |
| 7934  | MILLER     | CLERK     | 7782 | 1982-01-23 | 1300    | (null)   | 10 |
-------------------------------------------------------------------------------------------
```

图9.12 eval输出查询结果

示例 9-12

运行 Sqoop 脚本以封装 import 操作。

分析：

前面介绍的 Sqoop 的使用方式都是直接运行 Sqoop 脚本，这种方式使用起来比较麻烦。在 Sqoop 中提供了--options-file 参数，开发人员可以先将 Sqoop 脚本封装到一个文件中，然后使用--options-file 参数来指定封装后的脚本并运行，这样可以方便后期的维护。

关键代码：

```
//创建 emp.opt 文件，注意每个参数和值均须占单独一行。
$sqoop import \
--connect jdbc:mysql://master:3306/sqoop \
--username root \
--password 123456 \
--table emp \
--target-dir EMP_OPTIONS_FILE \
-m 2
//运行脚本
$ sqoop --options-file ./emp.opt
//查看运行结果
$ hdfs dfs -ls /user/root/EMP_OPTIONS_FILE
$ hdfs dfs -text /user/root/EMP_OPTIONS_FILE/part-m-*
```

上述代码运行结果如图 9.13 所示。

```
7369,SMITH,CLERK,7902,1980-12-17,800,null,20
7499,ALLEN,SALESMAN,7698,1981-02-20,1600,300,30
7521,WARD,SALESMAN,7698,1981-02-22,1250,500,30
7566,JONES,MANAGER,7839,1981-04-02,2975,null,20
7654,MARTIN,SALESMAN,7698,1981-09-28,1250,1400,30
7698,BLAKE,MANAGER,7839,1981-05-01,2850,null,30
7782,CLARK,MANAGER,7839,1981-06-09,2450,null,10
7788,SCOTT,ANALYST,7566,1987-07-13,3000,null,20
7839,KING,PRESIDENT,null,1981-11-17,5000,null,10
7844,TURNER,SALESMAN,7698,1981-09-08,1500,0,30
7876,ADAMS,CLERK,7788,1987-07-13,1100,null,20
7900,JAMES,CLERK,7698,1981-12-03,950,null,30
7902,FORD,ANALYST,7566,1981-12-03,3000,null,20
7934,MILLER,CLERK,7782,1982-01-23,1300,null,10
```

图9.13 运行sqoop脚本

9.1.3 导入 MySQL 数据到 Hive

将 MySQL 数据导入到 Hive 的执行原理：先将 MySQL 数据导入到 HDFS 上，然后再使用 load 函数将 HDFS 的文件加载到 Hive 表中。

示例 9-13

导入 sqoop.emp 表的所有字段到 Hive 中。

关键代码：

```
$ sqoop import \
--connect jdbc:mysql://localhost:3306/sqoop \
--username root --password 123456 \
--delete-target-dir \
--table emp \
--hive-import    --create-hive-table    --hive-table emp_import \
-m 1
```

注意

➢ 须使用--hive-import 来表示将数据导入到 Hive 中。

➢ 须使用--create-hive-table 来表示是否自动创建 hive 表；在实际编码过程中一般不使用该参数，而是通过手工事先将 hive 表创建好，这是因为自动创建的表的字段类型可能与理想中的表有差别。

➢ 须使用--hive-table 来指定要导入 Hive 的表的名称。

➢ 如果要导入到指定的 hive 数据库，可以通过--hive-database 参数来进行指定。

导入结果如图 9.14 所示。

```
hive> select * from default.emp_import;
OK
7369	SMITH	CLERK	7902	1980-12-17	800	NULL	20
7499	ALLEN	SALESMAN	7698	1981-02-20	1600	300	30
7521	WARD	SALESMAN	7698	1981-02-22	1250	500	30
7566	JONES	MANAGER	7839	1981-04-02	2975	NULL	20
7654	MARTIN	SALESMAN	7698	1981-09-28	1250	1400	30
7698	BLAKE	MANAGER	7839	1981-05-01	2850	NULL	30
7782	CLARK	MANAGER	7839	1981-06-09	2450	NULL	10
7788	SCOTT	ANALYST	7566	1987-07-13	3000	NULL	20
7839	KING	PRESIDENT	NULL	1981-11-17	5000	NULL	10
7844	TURNER	SALESMAN	7698	1981-09-08	1500	0	30
7876	ADAMS	CLERK	7788	1987-07-13	1100	NULL	20
7900	JAMES	CLERK	7698	1981-12-03	950	NULL	30
7902	FORD	ANALYST	7566	1981-12-03	3000	NULL	20
7934	MILLER	CLERK	7782	1982-01-23	1300	NULL	10
Time taken: 5.172 seconds, Fetched: 14 row(s)
```

图9.14 将MySQL数据导入到Hive

示例 9-14

导入表的指定字段到 Hive 中。

实现思路如下。

（1）在 Hive 中创建需要导入的 Hive 表。

```
hive> create table emp_column(
empno int,
ename string,
job string,
mgr int,
hiredate string,
sal double,
comm double,
deptno int
)
row format delimited fields terminated by '\t' lines terminated by '\n'
stored as textfile;
```

（2）使用 Sqoop 将 MySQL 表中指定的字段导入到 Hive 表中。

```
$ sqoop import \
--connect jdbc:mysql://master:3306/sqoop \
--username root --password 123456 \
--delete-target-dir \
--table emp \
--columns "EMPNO,ENAME,JOB,SAL,COMM" \
--fields-terminated-by '\t' --lines-terminated-by '\n' \
--hive-import --hive-overwrite --hive-table emp_column \
-m 1
```

上述代码中--hive-overwrite 表示覆盖已有数据。导入结果如图 9.15 所示。

```
hive> select * from emp_column;
OK
7369    SMITH   CLERK   800     null    NULL    NULL    NULL
7499    ALLEN   SALESMAN        1600    300     NULL    NULL    NULL
7521    WARD    SALESMAN        1250    500     NULL    NULL    NULL
7566    JONES   MANAGER 2975    null    NULL    NULL    NULL
7654    MARTIN  SALESMAN        1250    1400    NULL    NULL    NULL
7698    BLAKE   MANAGER 2850    null    NULL    NULL    NULL
7782    CLARK   MANAGER 2450    null    NULL    NULL    NULL
7788    SCOTT   ANALYST 3000    null    NULL    NULL    NULL
7839    KING    PRESIDENT       5000    null    NULL    NULL    NULL
7844    TURNER  SALESMAN        1500    0       NULL    NULL    NULL
7876    ADAMS   CLERK   1100    null    NULL    NULL    NULL
7900    JAMES   CLERK   950     null    NULL    NULL    NULL
7902    FORD    ANALYST 3000    null    NULL    NULL    NULL
7934    MILLER  CLERK   1300    null    NULL    NULL    NULL
```

图9.15 导入表的指定字段到Hive中

9.1.4 导入 MySQL 数据到 HBase

除了 HDFS 与 Hive 导入以外，Sqoop 还提供了 HBase 导入。通过指定“--hbase-table”可以导入 MySQL 的数据至 HBase 的一个表中。MySQL 表中的每一行记录将被转化为 HBase 的一个 Put 操作，操作结果为输出表中的一行。默认情况下，Sqoop 会将“--split-by”

列作为 RowKey，如果“--split-by”未指定，则会尝试选择主键列，也可能会通过“--hbase-row-key”来手动指定 RowKey。每一个输出列都会在一个相同的列族下由“--column-family”来指定。

示例 9-15

导入 MySQL 数据到 HBase。

关键代码：

```
$ sqoop import \
--connect jdbc:mysql://master:3306/sqoop \
--username root --password 123456 \
--table emp \
--columns "EMPNO,ENAME,JOB,SAL,COMM" \
--hbase-create-table \
--hbase-table emp_hbase_import \
--column-family details \
--hbase-row-key EMPNO \
-m 1
```

注意

- 须使用--hbase-create-table 来表示自动创建 HBase 表，表名由--hbase-table 来指定。
- --hbase-table 和--column-family 必须同时被指定。

导入结果如图 9.16 所示。

```
hbase(main):006:0> scan 'emp_hbase_import'
ROW                            COLUMN+CELL
 7369                          column=details:ENAME, timestamp=1547111295751, value=SMITH
 7369                          column=details:JOB, timestamp=1547111295751, value=CLERK
 7369                          column=details:SAL, timestamp=1547111295751, value=800
 7499                          column=details:COMM, timestamp=1547111295751, value=300
 7499                          column=details:ENAME, timestamp=1547111295751, value=ALLEN
 7499                          column=details:JOB, timestamp=1547111295751, value=SALESMAN
 7499                          column=details:SAL, timestamp=1547111295751, value=1600
 7521                          column=details:COMM, timestamp=1547111295751, value=500
 7521                          column=details:ENAME, timestamp=1547111295751, value=WARD
 7521                          column=details:JOB, timestamp=1547111295751, value=SALESMAN
 7521                          column=details:SAL, timestamp=1547111295751, value=1250
 7566                          column=details:ENAME, timestamp=1547111295751, value=JONES
```

图9.16 导入MySQL数据到HBase

9.1.5 导出 HDFS 数据到 MySQL

在导出数据（表）前需要先创建待导出表的结构，如果待导出的表在数据库中不存在，则报错；如果重复导出表，则表中的数据会重复。

1. MySQL 准备导出表

在 MySQL 数据库中创建要导出的表，并直接根据 emp 表创建导出表的结构。

```
mysql> create table emp_demo as select * from emp where 1=2;
```

2. sqoop-export

使用 sqoop 导出操作，开发人员可以通过 sqoop 的 help 命令查看该操作应该如何使用；重要参数的设置如下列代码段所示，本章之后的案例都是基于这些参数进行代码编写的。

```
sqoop help export
usage: sqoop export [GENERIC-ARGS] [TOOL-ARGS]
//  通用参数
Common arguments:
    --connect <jdbc-uri>
    --password <password>
    --username <username>
// 导出控制参数
Export control arguments:
    --batch
    --columns <col,col,col...>
    --direct
    --export-dir <dir>
    -m,--num-mappers <n>
    --mapreduce-job-name <name>
    --table <table-name>
// 输入文件参数配置
Input parsing arguments:
    --input-fields-terminated-by <char>
    --input-lines-terminated-by <char>
// 输出文件参数配置
Output line formatting arguments:
    --fields-terminated-by <char>
    --lines-terminated-by <char>
```

由上述代码可以发现，这些参数大部分与 import 参数类似，它们的不同之处将通过下列示例进行说明。

示例 9-16

导出 HDFS 中/user/root/emp 中的数据至 MySQL。

关键代码：

```
$ sqoop export \
--connect jdbc:mysql://master:3306/sqoop \
--username root --password 123456 \
--table emp_demo \
--export-dir /user/root/emp \
-m 1
```

须使用--export-dir 来指出将要导出的数据目录。

注意

每运行一次上述代码，就会重新插入数据到 MySQL 中，所以在运行代码时要先根据条件将表中的数据删除后再导出数据。

导出结果如图 9.17 所示。

```
mysql> select * from emp_demo;
+-------+--------+-----------+------+------------+------+------+--------+
| EMPNO | ENAME  | JOB       | MGR  | HIREDATE   | SAL  | COMM | DEPTNO |
+-------+--------+-----------+------+------------+------+------+--------+
|  7369 | SMITH  | CLERK     | 7902 | 1980-12-17 |  800 | NULL |     20 |
|  7499 | ALLEN  | SALESMAN  | 7698 | 1981-02-20 | 1600 |  300 |     30 |
|  7521 | WARD   | SALESMAN  | 7698 | 1981-02-22 | 1250 |  500 |     30 |
|  7566 | JONES  | MANAGER   | 7839 | 1981-04-02 | 2975 | NULL |     20 |
|  7654 | MARTIN | SALESMAN  | 7698 | 1981-09-28 | 1250 | 1400 |     30 |
|  7698 | BLAKE  | MANAGER   | 7839 | 1981-05-01 | 2850 | NULL |     30 |
|  7782 | CLARK  | MANAGER   | 7839 | 1981-06-09 | 2450 | NULL |     10 |
|  7788 | SCOTT  | ANALYST   | 7566 | 1987-07-13 | 3000 | NULL |     20 |
|  7839 | KING   | PRESIDENT | NULL | 1981-11-17 | 5000 | NULL |     10 |
|  7844 | TURNER | SALESMAN  | 7698 | 1981-09-08 | 1500 |    0 |     30 |
|  7876 | ADAMS  | CLERK     | 7788 | 1987-07-13 | 1100 | NULL |     20 |
|  7900 | JAMES  | CLERK     | 7698 | 1981-12-03 |  950 | NULL |     30 |
|  7902 | FORD   | ANALYST   | 7566 | 1981-12-03 | 3000 | NULL |     20 |
|  7934 | MILLER | CLERK     | 7782 | 1982-01-23 | 1300 | NULL |     10 |
+-------+--------+-----------+------+------------+------+------+--------+
```

图9.17　HDFS数据导出至MySQL

示例 9-17

导出表的指定字段。

关键代码：

```
$ sqoop export \
--connect jdbc:mysql://master:3306/sqoop \
--username root --password 123456 \
--table emp_demo \
--columns "EMPNO, ENAME, JOB, SAL, COMM" \
--export-dir /user/root/emp_column   \
-m 1
```

注意

为了成功实现结果测试，建议先删除目标表中的数据（DELETE FROM emp_demo）。

导出结果如图 9.18 所示。

示例 9-18

导出表时指定分隔符。

关键代码：

```
$ sqoop export \
--connect jdbc:mysql://master:3306/sqoop \
--username root --password 123456 \
```

```
--table emp_demo \
--columns "EMPNO, ENAME, JOB, SAL, COMM" \
--export-dir /user/root/emp_column_split  \
--fields-terminated-by '\t' --lines-terminated-by '\n' \
-m 1
```

EMPNO	ENAME	JOB	MGR	HIREDATE	SAL	COMM	DEPTNO
7369	SMITH	CLERK	NULL	NULL	800	NULL	NULL
7499	ALLEN	SALESMAN	NULL	NULL	1600	300	NULL
7521	WARD	SALESMAN	NULL	NULL	1250	500	NULL
7566	JONES	MANAGER	NULL	NULL	2975	NULL	NULL
7654	MARTIN	SALESMAN	NULL	NULL	1250	1400	NULL
7698	BLAKE	MANAGER	NULL	NULL	2850	NULL	NULL
7782	CLARK	MANAGER	NULL	NULL	2450	NULL	NULL
7788	SCOTT	ANALYST	NULL	NULL	3000	NULL	NULL
7839	KING	PRESIDENT	NULL	NULL	5000	NULL	NULL
7844	TURNER	SALESMAN	NULL	NULL	1500	0	NULL
7876	ADAMS	CLERK	NULL	NULL	1100	NULL	NULL
7900	JAMES	CLERK	NULL	NULL	950	NULL	NULL
7902	FORD	ANALYST	NULL	NULL	3000	NULL	NULL
7934	MILLER	CLERK	NULL	NULL	1300	NULL	NULL

图9.18　导出表的指定字段至MySQL

注意

须使用--fields-terminated-by 和--lines-terminated-by 参数来指定数据行和列的分隔符。

导出结果如图 9.19 所示。

EMPNO	ENAME	JOB	MGR	HIREDATE	SAL	COMM	DEPTNO
7369	SMITH	CLERK	NULL	NULL	800	NULL	NULL
7499	ALLEN	SALESMAN	NULL	NULL	1600	300	NULL
7521	WARD	SALESMAN	NULL	NULL	1250	500	NULL
7566	JONES	MANAGER	NULL	NULL	2975	NULL	NULL
7654	MARTIN	SALESMAN	NULL	NULL	1250	1400	NULL
7698	BLAKE	MANAGER	NULL	NULL	2850	NULL	NULL
7782	CLARK	MANAGER	NULL	NULL	2450	NULL	NULL
7788	SCOTT	ANALYST	NULL	NULL	3000	NULL	NULL
7839	KING	PRESIDENT	NULL	NULL	5000	NULL	NULL
7844	TURNER	SALESMAN	NULL	NULL	1500	0	NULL
7876	ADAMS	CLERK	NULL	NULL	1100	NULL	NULL
7900	JAMES	CLERK	NULL	NULL	950	NULL	NULL
7902	FORD	ANALYST	NULL	NULL	3000	NULL	NULL
7934	MILLER	CLERK	NULL	NULL	1300	NULL	NULL

图9.19　导出表时指定分隔符

示例 9-19

批量导出数据。

关键代码：

```
$ sqoop export \
```

```
-Dsqoop.export.records.pre.statement=10 \
--connect jdbc:mysql://master:3306/sqoop \
--username root --password 123456 \
--table emp_demo \
--export-dir /user/root/emp
```

注意

➢ 默认情况下，读取 HDFS 文件的一行数据后会插入一条记录到关系型数据库中，性能低下。

➢ 批量导出的数据须使用参数-Dsqoop.export.records.pre.statement 来指定，可一次导出指定数据到关系型数据库中。

导出结果如图 9.20 所示。

```
+-------+--------+-----------+------+------------+------+------+--------+
| EMPNO | ENAME  | JOB       | MGR  | HIREDATE   | SAL  | COMM | DEPTNO |
+-------+--------+-----------+------+------------+------+------+--------+
|  7900 | JAMES  | CLERK     | 7698 | 1981-12-03 |  950 | NULL |     30 |
|  7902 | FORD   | ANALYST   | 7566 | 1981-12-03 | 3000 | NULL |     20 |
|  7934 | MILLER | CLERK     | 7782 | 1982-01-23 | 1300 | NULL |     10 |
|  7369 | SMITH  | CLERK     | 7902 | 1980-12-17 |  800 | NULL |     20 |
|  7499 | ALLEN  | SALESMAN  | 7698 | 1981-02-20 | 1600 |  300 |     30 |
|  7521 | WARD   | SALESMAN  | 7698 | 1981-02-22 | 1250 |  500 |     30 |
|  7566 | JONES  | MANAGER   | 7839 | 1981-04-02 | 2975 | NULL |     20 |
|  7654 | MARTIN | SALESMAN  | 7698 | 1981-09-28 | 1250 | 1400 |     30 |
|  7698 | BLAKE  | MANAGER   | 7839 | 1981-05-01 | 2850 | NULL |     30 |
|  7782 | CLARK  | MANAGER   | 7839 | 1981-06-09 | 2450 | NULL |     10 |
|  7788 | SCOTT  | ANALYST   | 7566 | 1987-07-13 | 3000 | NULL |     20 |
|  7839 | KING   | PRESIDENT | NULL | 1981-11-17 | 5000 | NULL |     10 |
|  7844 | TURNER | SALESMAN  | 7698 | 1981-09-08 | 1500 |    0 |     30 |
|  7876 | ADAMS  | CLERK     | 7788 | 1987-07-13 | 1100 | NULL |     20 |
+-------+--------+-----------+------+------------+------+------+--------+
```

图9.20　批量导出数据

9.1.6　技能实训

请将 Hive 中零售商店数据导出至 MySQL 中。

关键步骤：

➢ 导出顾客表；

➢ 导出订单表；

➢ 导出订单明细表；

➢ 导出商品表。

任务 2　使用 Sqoop Job 完成 Hive 与 MySQL 间的数据迁移

【任务描述】

通过 Sqoop job 实现 Hive 与 MySQL 间的数据迁移。

【关键步骤】

（1）使用 Sqoop job 实现 MySQL 数据导入 Hive。

（2）使用 Sqoop job 实现 Hive 数据导出至 MySQL。

9.2.1 Sqoop Job

Sqoop Job 工具允许开发人员将 Sqoop 脚本创建为作业，这样方便维护与重复调用。

【语法】

```
sqoop job (generic-args) (job-args) [-- [subtool-name] (subtool-args)]
```

Sqoop 作业管理命令参数如表 9-7 所示。

表 9-7 Sqoop 作业管理命令参数

参数	描述
--create <job-id>	创建作业
--delete <job-id>	删除作业
--exec <job-id>	执行作业
--show <job-id>	查看作业参数信息
--list	列出所有作业

1. 创建作业

示例 9-20

定义 Sqoop Job 以实现 MySQL 数据导入 Hive。

关键代码：

```
$ sqoop job --create myjob -- \
import --connect jdbc:mysql://master:3306/sqoop \
--username root \
--delete-target-dir \
--table emp
```

查看作业：

```
$ sqoop job --show myjob
```

作业执行结果如图 9.21 所示。

```
Enter password:
Job: myjob
Tool: import
Options:
----------------------------
reset.onemapper = false
codegen.output.delimiters.enclose = 0
sqlconnection.metadata.transaction.isolation.level = 2
codegen.input.delimiters.escape = 0
codegen.auto.compile.dir = true
accumulo.batch.size = 10240000
```

图9.21 Sqoop作业执行结果

注意

在查看作业时若提示输入密码，请输入 MySQL 数据库的密码。

2. **执行作业**

示例 9-21

执行示例 9-20 中创建的作业。

关键代码：

```
$ sqoop job --exec myjob
```

执行结果与示例 9-5 相同。

9.2.2 技能实训

使用 Sqoop Job 完成 Hive 数据导出至 MySQL。

关键步骤：

（1）创建 Sqoop Job；

（2）编写 Sqoop export 脚本；

（3）执行 Sqoop Job。

本章小结

➢ Sqoop 是数据迁移工具，通常被应用在 RDBMS 与 Hadoop 之间的数据迁移场景中。

➢ sqoop-import 是数据导入工具，用于将 RDBMS 中的数据导入至 Hadoop 中。

➢ sqoop-export 是数据导出工具，用于将 Hadoop 中的数据导出至 MySQL 中。

➢ Sqoop Job 是作业管理工具，可以将 Sqoop 脚本封装为作业，以方便维护与调用。

本章作业

一、简答题

1. Sqoop 是什么？

2. Sqoop 常用工具包括哪些？

二、编码题

1. 使用 Sqoop 以增量的方式导入数据，要求只导入 empno>7788 的数据到 HDFS 中。关于 Sqoop 增量导入的介绍请扫描二维码获取。

2．将本章中的所有示例封装为脚本，并使用“--options-file”参数对它们进行调用。

3．将本章中的所有导入/导出示例封装为作业，并使用“sqoop job --exec”命令执行之。

Sqoop 增量导入

第 10 章

项目实训：电子商务消费行为分析

技能目标

- 熟练掌握 Hive 数据定义（DDL）操作。
- 熟练掌握 Hive 数据装载（DML）操作。
- 熟练掌握 Hive 数据查询操作。
- 熟练掌握 Hive 视图操作。
- 熟练掌握聚合函数操作。
- 熟练掌握窗口函数操作。

本章任务

根据某零售企业近一年收集的数据完成“电子商务消费行为分析”项目实训。

本章资源下载

第10章 项目实训：电子商务消费行为分析
- 10.1 项目准备
- 10.2 难点分析
- 10.3 项目实现思路

10.1 项目准备

1. 项目需求分析

对某零售企业的门店最近一年收集的数据进行分析。

- 潜在客户分析并画像。
- 用户消费情况统计。
- 门店资源利用率分析。
- 消费的特征人群定位。
- 数据可视化展现。

数据包括客户信息（customer_details.csv，数据结构描述如表 10-1 所示）、交易信息（transaction_details.csv，数据结构描述如表 10-2 所示）、门店信息（store_details.csv，数据结构描述如表 10-3 所示）以及评价信息（store_review.csv，数据结构描述如表 10-4 所示）。

表 10-1　客户信息数据结构描述

字段	类型	说明
customer_id	int	客户编号
first_name	string	名
last_name	string	姓
email	string	电子邮箱
gender	string	性别
address	string	联系地址
country	string	国家地区
language	string	语言
job	string	职业
credit_type	string	信用卡类型
credit_no	string	信用卡号

表 10-2　交易信息数据结构描述

字段	类型	说明
transaction_id	int	交易流水号
customer_id	int	客户编号
store_id	int	门店编号

续表

字段	类型	说明
price	decimal	交易金额，如 5.08
product	string	交易商品
date	string	交易日期
time	string	交易时间

表 10-3　门店信息数据结构描述

字段	类型	说明
store_id	int	门店编号
store_name	string	门店名称
employee_number	int	雇员数量

表 10-4　评价信息数据结构描述

字段	类型	说明
transaction_id	int	交易流水号
store_id	int	门店编号
review_store	int	评分

2. 项目环境准备

完成某企业“电子商务消费数据分析”项目实训，须具备的开发环境要求如下。

- 开发环境：Hadoop，Hive。
- 开发工具：Zeppelin。

本项目中所有操作均采用 Zeppelin 来完成，所以需要先启动 HiveServer2 以连接 Zeppelin。前面简单介绍过 Zeppelin 的使用方法，这里再强调一下安装 Zeppelin 的几个关键步骤。

（1）下载 Zeppelin 安装包，本书采用的是目前最新版本的 Zeppelin（zeppelin-0.8.0-bin-all.taz，该版本集成了所有解释器）。

（2）Zeppelin 安装包解压后，将 Hive 相关的 JAR 文件（如下所示）复制到 $ZEPPELIN_HOME/lib 目录下。

- $HIVE_HOME/lib/hive-common-1.1.0-cdh5.14.2.jar
- $HIVE_HOME/lib/hive-jdbc-1.1.0-cdh5.14.2.jar
- $HIVE_HOME/lib/hive-service-1.1.0-cdh5.14.2.jar

（3）启动 Zeppelin，代码如下。

```
$ZEPPELIN_HOME/bin/zeppelin-daemon.sh start
```

使用 Web 地址进入 Zeppelin，地址中会出现的 HOSTNAME 为 Zeppelin 安装所在主机名。

（4）向 Zeppelin 中添加“Hive”解释器，添加步骤如图 10.1 至图 10.3 所示。

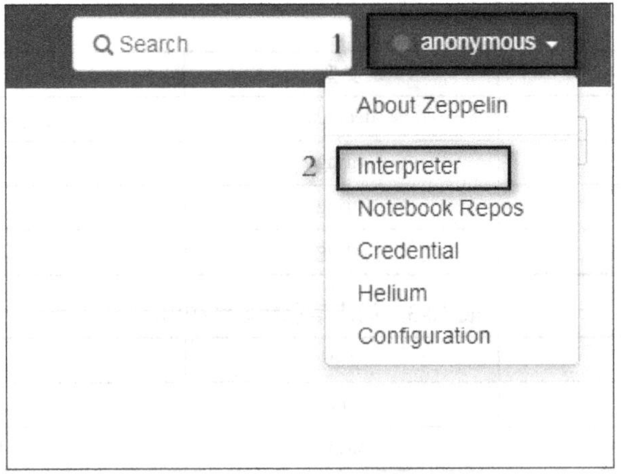

图10.1　添加Hive解释器步骤1

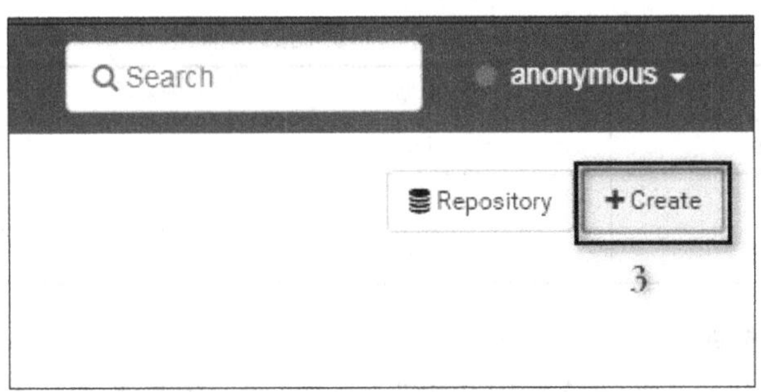

图10.2　添加Hive解释器步骤2

图10.3　添加Hive解释器步骤3

（5）“Notebook”→“Create new note”，输入如下语句测试“Hive”解释器。

```
%hive
show databases
```

一次运行一条语句，语句末尾无须以“;”结束。正确的运行结果如图 10.4 所示。

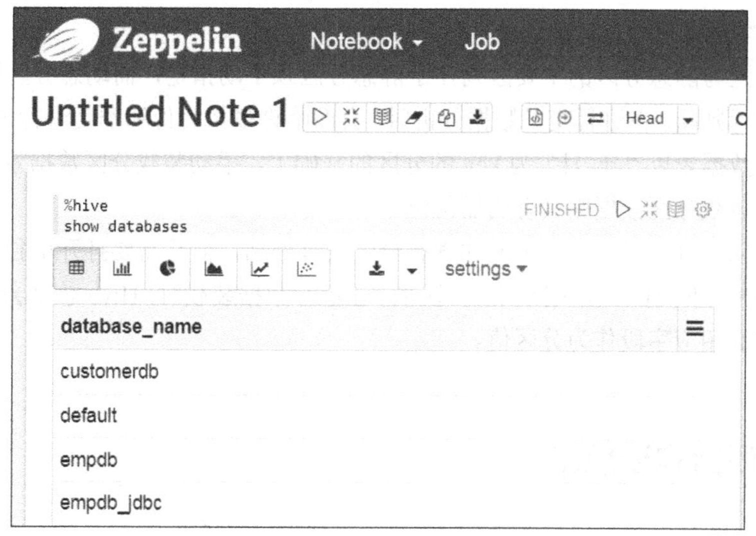

图10.4　新建notebook代码运行结果

3. ***项目覆盖的技能点***

项目覆盖的技能点包括：

- Hive 数据定义（DDL）操作；
- Hive 数据装载（DML）操作；
- Hive 数据查询操作；
- Hive 视图操作；
- 聚合函数操作；
- 窗口函数操作；
- Zeppelin 应用。

10.2 难点分析

1. ***数据清洗***

收集的数据总会存在一些问题，常见的问题有：缺失值、异常值、重复值。

（1）缺失值的处理

如果缺失率较高，一般可以考虑直接删除该字段或属性。如果缺失率较低，可以考虑使用多种方法对缺失值进行填充，填充方法具体包括均值填充、中位数填充、常量填充等。

（2）异常值的处理

通常使用的处理方法包括：删除异常值、用平均值替代或将其视为缺失值。

（3）重复值的处理

在 Hive 中有 3 种除去重复值的办法：DISTINCT、GROUP BY 以及 ROW_NUMBER。

2. **动态分区**

动态分区与静态分区的主要区别在于静态分区须手动指定，而动态分区则须通过数据来对其进行判断。在进行交易数据分析时，会面临许多按月份和季度进行统计的需求，在完成交易数据表以“年-月”为 key 的分区的基础上，手动装载分区数据效率低下，相反动态自动装载分区数据则具有明显优势。

在使用动态分区时，表的定义和静态分区没有差别，但是需要打开单独的动态分区开关，默认情况下，Hive 是关闭动态分区的。在进行数据装载时，Hive 总是会将 SELECT 子句最末尾的相应字段作为分区值。

10.3 项目实现思路

1. **数据探索**

数据探索的目的是理解数据结构，了解数据规模，并尝试初步找出数据中的一些问题。

（1）下载 4 个数据文件至指定目录，如“~/workspace/hive/store/”，数据文件请扫描二维码获取。

数据文件

（2）检查总行数和 CSV 文件头。

关键代码：

```
%sh
cd ~/workspace/hive/store/
wc -l customer_details.csv
wc -l store_details.csv
wc -l store_review.csv
wc -l transaction_details.csv
head -2 customer_details.csv
head -2 store_details.csv
head -2 store_review.csv
head -2 transaction_details.csv
```

代码运行结果结果如图 10.5 所示。

```
501 customer_details.csv
6 store_details.csv
1001 store_review.csv
8101 transaction_details.csv
customer_id,first_name,last_name,email,gender,address,country,
1,Spencer,Raffeorty,sraffeorty0@dropbox.com,Male,9274 Lyons Co
store_id,store_name,employee_number
1,NoFrill,10
transaction_id,store_id,review_score
7430,1,5
transaction_id,customer_id,store_id,price,product,date,time
1,225,5,47.02,Bamboo Shoots - Sliced,2017-08-04,8:18
```

图10.5　数据探索

注意

如果未能显示文件内容，请使用 dos2unix 命令对 CSV 文件进行转换。

经过数据探索，发现已知的数据存在如下问题。

➢ 在“customer_details.csv”中，“language”字段存在编码错误，如“148”号客户这一列出现乱码。

➢ 在“transaction_details.csv”中，“transaction_id”字段中存在数据重复问题，如存在两个“8001”和两个“8002”等。这些数据是交易 id，具有唯一性，但是因为这些交易数据来自不同的门店，故发生了重复问题，之后应对这些重复数据进行处理。

➢ 在“store_review.csv”中，“review_score”存在为 NULL 的情况，并且有些“transaction_id”（如 transaction_id=912）映射到了错误的“store_id”，这些数据是无效的，应该被剔除。

上述问题均可在数据清洗阶段被解决。

2. 数据收集

将下载的数据文件上传至 HDFS。

关键代码：

```
%sh
cd ~/workspace/hive/store/
hdfs dfs -rm -r -f –skipTrash /data/shopping/
hdfs dfs -mkdir -p /data/shopping/customer/
hdfs dfs -put customer_details.csv /data/shopping/customer/
hdfs dfs -mkdir -p /data/shopping/transaction/
hdfs dfs -put transaction _details.csv /data/shopping/transaction/
hdfs dfs -mkdir -p /data/shopping/store/
hdfs dfs -put store _details.csv /data/shopping/store/
hdfs dfs -mkdir -p /data/shopping/review/
hdfs dfs -put store_ review.csv /data/shopping/review/
hdfs dfs -ls -R /data/shopping/
```

代码运行结果如图 10.6 所示。

```
drwxr-xr-x   - root supergroup          0 2019-01-03 17:00 /data/shopping/customer
-rw-r--r--   3 root supergroup      63187 2019-01-03 17:00 /data/shopping/customer/customer_details.csv
drwxr-xr-x   - root supergroup          0 2019-01-03 17:00 /data/shopping/review
-rw-r--r--   3 root supergroup       8832 2019-01-03 17:00 /data/shopping/review/store_review.csv
drwxr-xr-x   - root supergroup          0 2019-01-03 17:00 /data/shopping/store
-rw-r--r--   3 root supergroup        105 2019-01-03 17:00 /data/shopping/store/store_details.csv
drwxr-xr-x   - root supergroup          0 2019-01-03 17:00 /data/shopping/transaction
-rw-r--r--   3 root supergroup     462327 2019-01-03 17:00 /data/shopping/transaction/transaction_details.csv
```

图10.6　数据收集

3. 外部表创建

原始数据文件很有可能被多个部门同时操作，此时须使用外部表进行建表，并由外

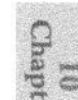

部表产生视图或生成新的内部分区表。需要说明的是，即使将元数据删除，HDFS 上的实际数据也不受影响。

对每一个 CSV 文件创建外部表。

关键代码：

```
%hive
create database if not exists shopping
use shopping
create external table if not exists ext_customer_details(
      customer_id string,
      first_name string,
      last_name string,
      email string,
      gender string,
      address string,
      country string,
      language string,
      job string,
      credit_type string,
      credit_no string
)
row format serde 'org.apache.hadoop.hive.serde2.OpenCSVSerde'
location '/data/shopping/customer'
tblproperties("skip.header.line.count"="1")
```

上述代码完成了数据库和客户信息表的创建，其他三个表（交易表 ext_transaction_details、门店表 ext_store_details 及评价表 ext_store_review）的创建可以参考上述代码完成，最后须对上述四个表的创建结果进行正确性验证。

```
%hive
select * from ext_customer_details limit 20
```

4. 数据清洗

（1）客户数据脱敏

客户信息中存在大量隐私信息，如客户名、邮件、联系地址及信用卡号等，应该对其进行加密。另外“language”字段还存在乱码情况，但是该字段对后续统计分析没有指导意义，故可将其删除。综上所述可知在 ext_customer_details 表的基础上可创建如下视图。

关键代码：

```
%hive
create view if not exists vw_customer_details as select
customer_id,
first_name,
unbase64(last_name) as last_name,
unbase64(email) as email,
gender,
```

```
unbase64(address) as address,
country,
job,
credit_type,
unbase64(concat(unbase64(credit_no),'seed')) as credit_no
from ext_customer_details
```

上述代码使用了内置函数 unbase64()，该函数可将 64 位字符串转换为二进制值。同时，新创建的视图 vw_customer_details 中也不再包含“language”列。

（2）清洗交易数据

首先，考虑到交易数据是整个业务系统数据中量最大的部分，并且明显按时间分布，同时也是后续数据分析的主要依据，所以可以将交易数据表进行分区，分区粒度此处选择按月份。

关键代码：

```
%hive
create table if not exists transaction_details(
transaction_id string,customer_id string,store_id string,price decimal(8,2),product string,
purchase_date date,purchase_time string)
partitioned by (purchase_month string)
```

其次，需要解决 ext_transaction_details 表中“transaction_id”重复的问题。因为这些交易编号来自于不同的门店，所以不能将其简单删除。考虑到这些编号只是一个标识，不对该列进行求和或求平均值，所以解决该问题的常用方法是：为重复的编号进行重新编号。最直接的方法是使用窗口函数 row_number()并按“transaction_id”进行分组以求出每条记录的组内序号，再组合原始交易编号进而形成新的交易编号。

关键代码：

```
%hive
--打开动态分区模式
set hive.exec.dynamic.partition.mode=nonstrict
--使用动态分区插入数据
with base as (
select transaction_id,customer_id,store_id,price,product,purchase_date,purchase_time,
from_uninxtime(unix_timestamp(purchase_date,'yyyy-MM-dd'),'yyyy-MM') as purchase_month,
row_number() over(partition by transaction_id order by store_id) as rn
from ext_transaction_details
where customer_id<>'customer_id'
)
from base
insert overwrite table transaction_details partition(purchase_month)
select
if(rn=1,transaction_id,concat(transaction_id,'_fix',rn)) as transaction_id,
customer_id,store_id,price,product,purchase_date,  purchase_time,purchase_month
--查看清洗效果
select transaction_id,customer_id,store_id,price,product,purchase_date,purchase_time,purchase_month
```

```
from transaction_details where transaction_id like '%fix%'
```

代码运行结果如图 10.7 所示。

transaction_id	customer_id	store_id	price
8011_fix2	190	5	36.7
8019_fix2	453	4	30.55

图10.7　清洗交易数据

（3）清洗评价数据

“transaction_id”与“store_id”在评价表 ext_store_review 与交易表中存在不一致的情况，可通过下述代码验证之。

```
%hive
select count(*) from ext_store_review r join ext_transaction_details t on
r.transaction_id=t.transaction_id and r.store_id=t.store_id
where review_score<>''
```

代码运行结果显示共有 153 行评价记录与交易记录匹配。

```
%hive
select count(*) from ext_store_review where review_score<>''
```

代码运行结果显示实际评价记录有 937 行，很明显评价表中“store_id”的错误率太高，所以可以创建新的评价视图以忽略“store_id”。

关键代码：

```
%hive
create view if not exists vw_store_review as
select transaction_id,review_score from ext_store_review where review_score<>''
```

5．***数据分析与可视化***

（1）客户分析

① 最受客户欢迎的信用卡

对 vw_customer_details 中每个客户的信用卡类型进行分组统计总数。考虑到客户可能来自不同国家，所以须对客户所在国家和持有的信用卡类型进行分组。

关键代码：

```
%hive
select credit_type,count(distinct credit_no) as credit_cnt
from vw_customer_details group by country,credit_type order by credit_cnt desc
```

分析结果以“饼图”（Pie Chart）输出显示，如图 10.8 所示。

② 排名前 5 的客户职业

对 vw_customer_details 中每个客户的职业进行分组统计总数。

关键代码：

```
%hive
select job,count(*) as pn from vw_customer_details group by job order by pn desc limit 5
```

统计结果以“柱状图”（Bar Chart）输出显示，如图 10.9 所示。

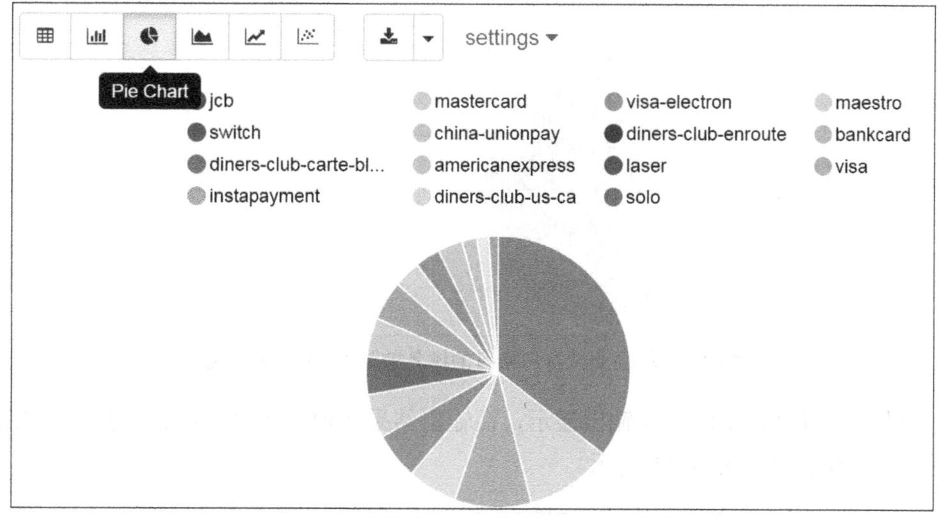

图10.8　最受客户欢迎的信用卡分析结果

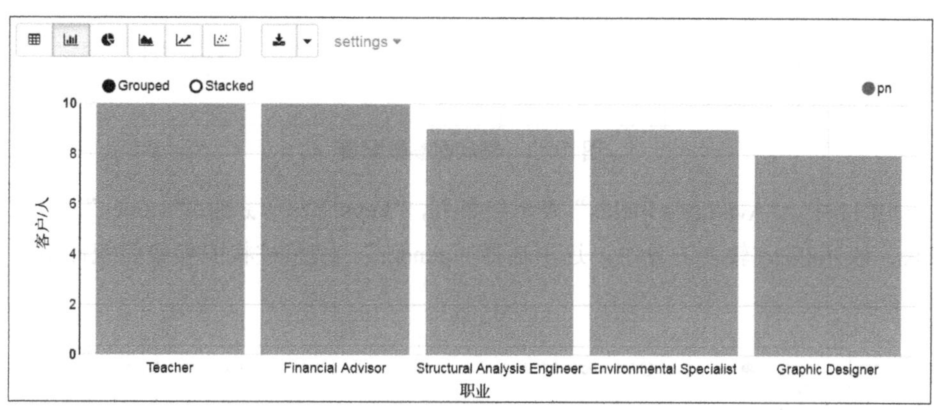

图10.9　排名前5的客户职业统计结果

③ 美国女性客户持有的排名前 3 的信用卡

首先对 vw_customer_details 中的客户进行过滤，要求客户的国家为“United States”，性别是“Female”；然后对过滤结果（美国女性客户）按信用卡类型进行分组统计总数。

关键代码：

```
%hvie
select credit_type,count(*) as ct from vw_customer_details
where country='United States' and gender='Female'
group by credit_type order by ct desc limit 3
```

统计结果以饼图输出显示，如图 10.10 所示。

④ 按国家和性别进行客户统计

对 vw_customer_details 中每个客户，按国家和性别分组统计总数。

关键代码：

```
%hive
select count(*),country,gender from vw_customer_details group by country,gender
```

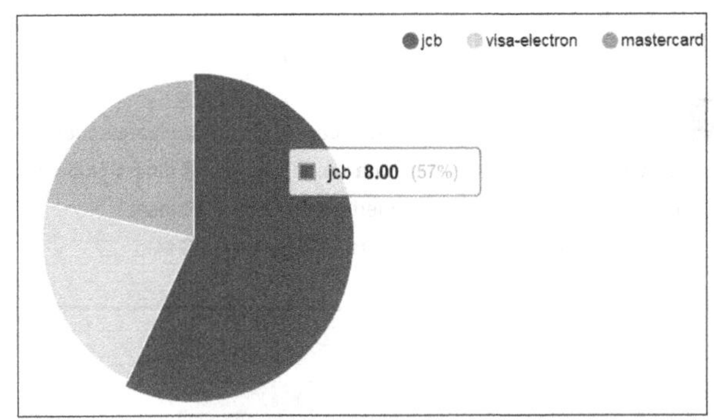

图10.10 美国女性客户持有的排名前3的信用卡统计结果

统计结果可以柱状图形式输出显示，在输出结果之前使用“settings”打开参数配置界面，其参数配置如图 10.11 所示。

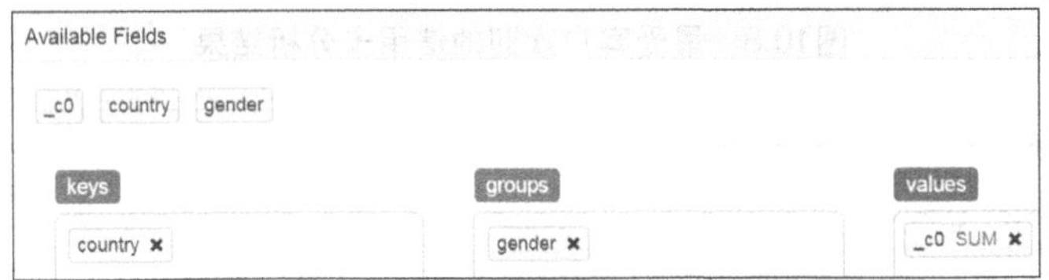

图10.11 柱状图参数配置

图 10.11 中，“Available Fields”表示结果列，“keys”表示 x 轴，“values”表示 y 轴，“groups”表示在 x 轴上的分组，这里选择“gender”。调整柱状图参数后统计结果如图 10.12 所示。

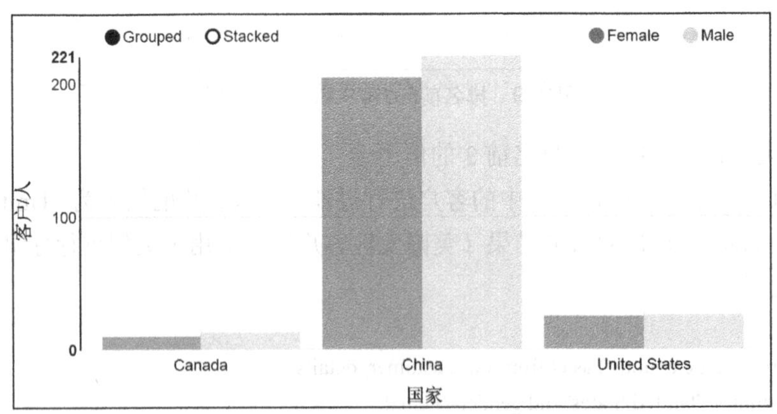

图10.12 按国家和性别进行客户统计结果

当然也可以将“keys”设为“gender”、将“groups”设为“country”，更多柱状图显示技巧可自行探索并掌握，更多可视化图表的配置请扫描二维码获取。

Zeppelin 可视化图表的配置

（2）交易分析

① 按月统计总收益

对 transaction_details 进行按月分组，并对交易金额求和。

关键代码：

```
%hive
select sum(price) as revenue_mom,purchase_month from transaction_details group by purchase_month
```

代码运行结果如图 10.13 所示。

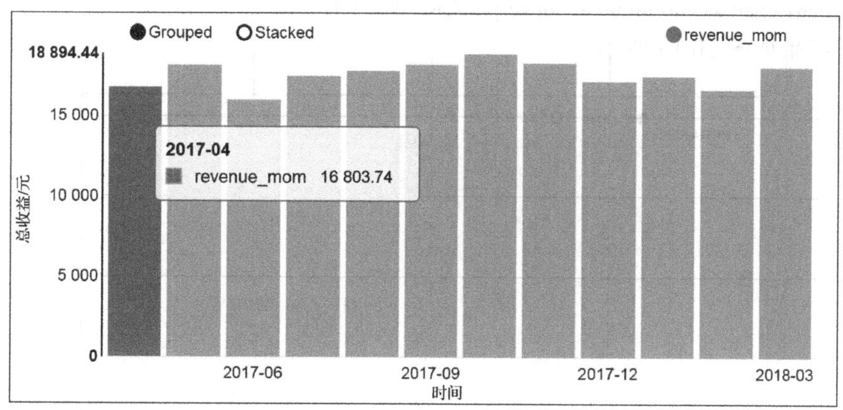

图10.13　按月统计总收益

② 按季度统计总收益

在“transaction_details”中没有季度字段，但可以通过日期字段“purchase_date”将“price”计算出来，然后按各季度对“price”分组求和。

关键代码：

```
%hive
with base as (select price,
concat_ws('-',substr(purchase_date,1,4),cast(ceil(month(purchase_date)/3.0) as string)) as year_quarter
from transaction_details)
select sum(price) as revenue_qoq,year_quarter from base group by year_quarter
```

统计结果如图 10.14 所示。

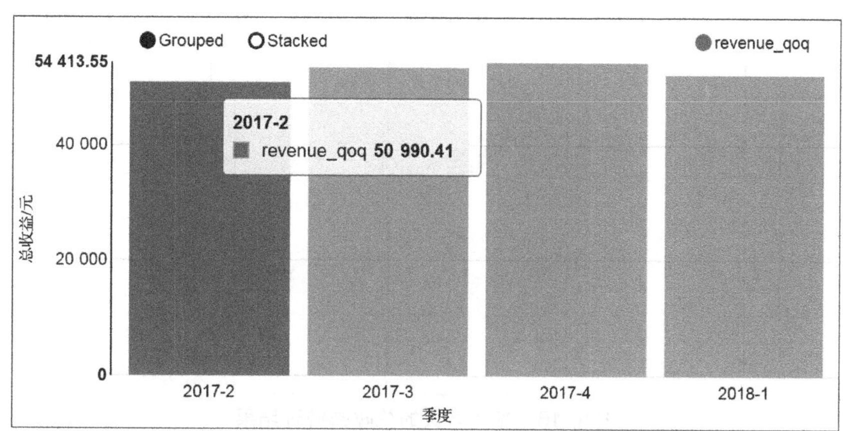

图10.14　按季度统计总收益

③ 按年统计总收益

对 transaction_details 进行按年分组，并对交易金额进行求和，注意年份需要单独计算出来。

关键代码：

```
%hive
select sum(price) as revenue_mom,substr(purchase_date,1,4) as year
from transaction_details group by substr(purchase_date,1,4)
```

统计结果如图 10.15 所示。

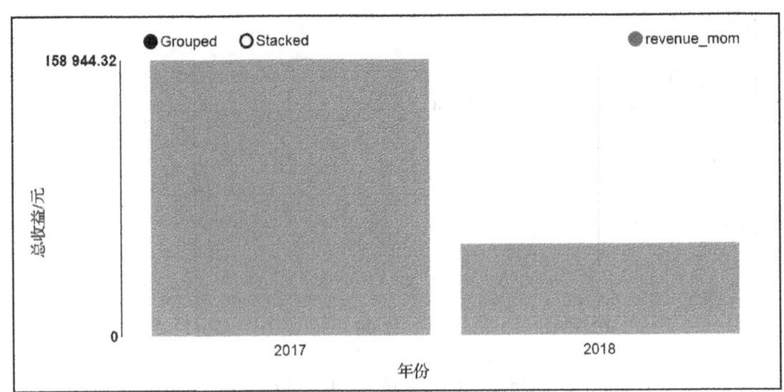

图10.15　按年统计总收益

④ 统计每周各天的总收益

统计每周各天的总收益，须首先计算出所有交易日期分别属于星期几，然后按星期进行分组并对“price”求和。

关键代码：

```
%hive
select sum(price) as revenue_wow,date_format(purchase_date,'u') as weekday
from transaction_details group by date_format(purchase_date,'u')
```

统计结果以饼图显示，设置“keys”为“weekday”，“values”为“revenue_wow”，则统计结果如图 10.16 所示。

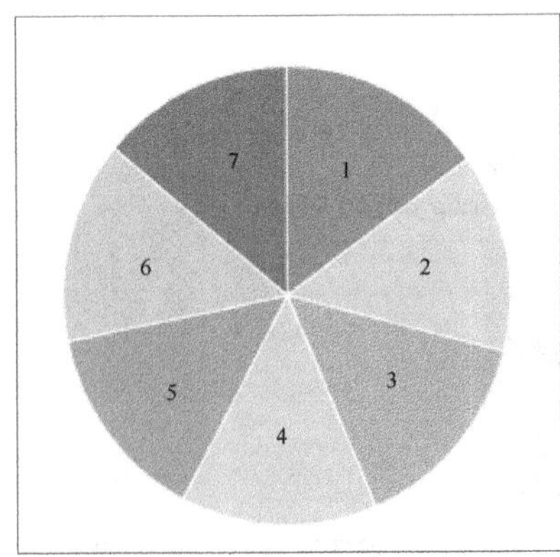

图10.16　每周各天的总收益统计结果

⑤ 按时间段统计平均收益（每次）和总收益

关键代码：

```
%hive
```

```
with base as (
select price, purchase_time, if(purchase_time like '%PM',
concat_ws(':',string(hour(from_unixtime(unix_timestamp(purchase_time,'hh:mm')))+12),
string(minute(from_unixtime(unix_timestamp(purchase_time,'hh:mm'))))),
from_unixtime(unix_timestamp(purchase_time,'hh:mm'),'HH:mm')) as time_format
from transaction_details
),
timeformat as (
select
purchase_time, price,
(cast(split(time_format, ':')[0] as decimal(4,2)) +    cast(split(time_format, ':')[1] as decimal(4,2))/60)
as purchase_time_in_hrs
from base
),
timebucket as (
select
price, purchase_time, purchase_time_in_hrs,
if(purchase_time_in_hrs > 5 and purchase_time_in_hrs <=8, 'early morning',
if(purchase_time_in_hrs > 8 and purchase_time_in_hrs <=11, 'morning',
if(purchase_time_in_hrs > 11 and purchase_time_in_hrs <=13, 'noon',
if(purchase_time_in_hrs > 13 and purchase_time_in_hrs <=18, 'afternoon',
if(purchase_time_in_hrs > 18 and purchase_time_in_hrs <=22, 'evening', 'night'))))) as time_bucket
from timeformat
)
select time_bucket, avg(price) as avg_spend, sum(price)/1000 as revenue_k
from timebucket group by time_bucket -- divide 1k to see the chater more clear
```

统计结果如图 10.17 所示。

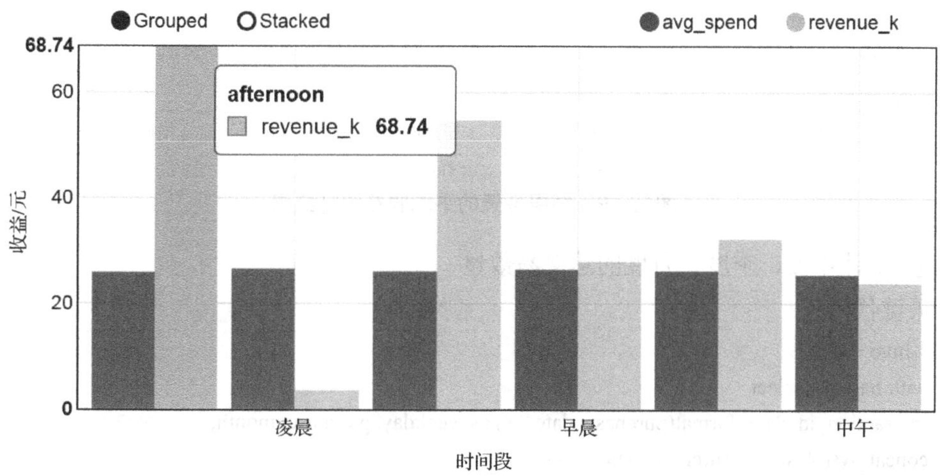

图10.17　按时间段统计平均收益（每次）和总收益

⑥ 统计每周各天的平均收益

关键代码：

```
%hive
select avg(price) as avg_price, date_format(purchase_date, 'u') as weekday from transaction_details
where date_format(purchase_date, 'u') is not null group by date_format(purchase_date, 'u')
```

拆线图参数配置如图 10.18 所示。

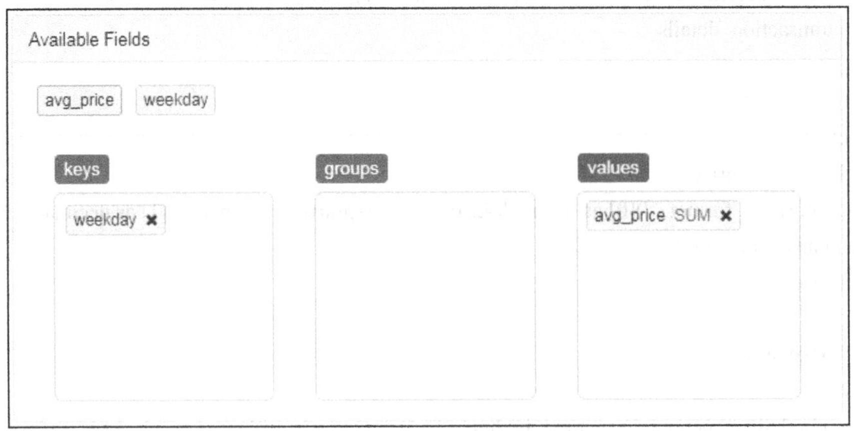

图10.18　拆线图参数配置

统计结果如图 10.19 所示。

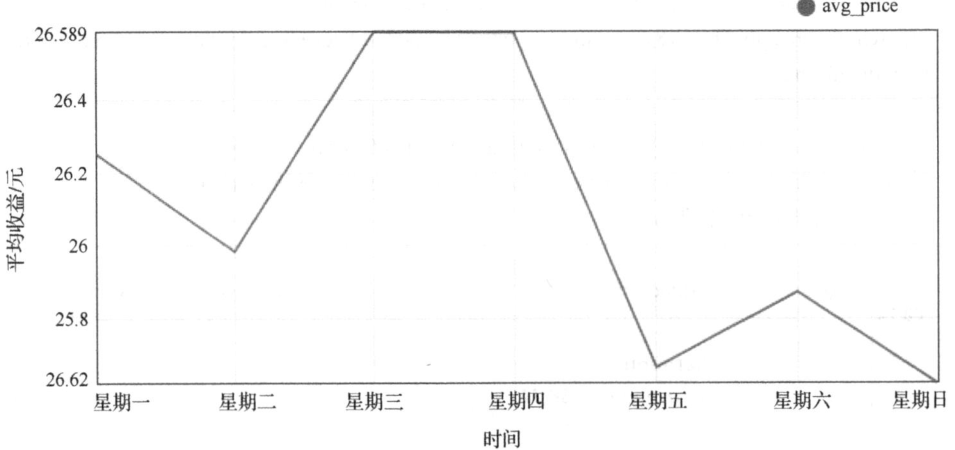

图10.19　每周各天的平均收益统计结果

⑦ 统计年度、季度、月度的总交易数量

关键代码：

```
%hive
with base as (select
transaction_id, date_format(purchase_date, 'u') as weekday, purchase_month,
concat_ws('-', substr(purchase_date, 1, 4),
cast(ceil(month(purchase_date)/3.0) as string)) as year_quarter, substr(purchase_date, 1, 4) as year
from transaction_details where purchase_month is not null)
select count(distinct transaction_id) as total, weekday, purchase_month, year_quarter, year
from base group by weekday, purchase_month, year_quarter, year order by year, purchase_month
```

统计结果如图 10.20 所示。

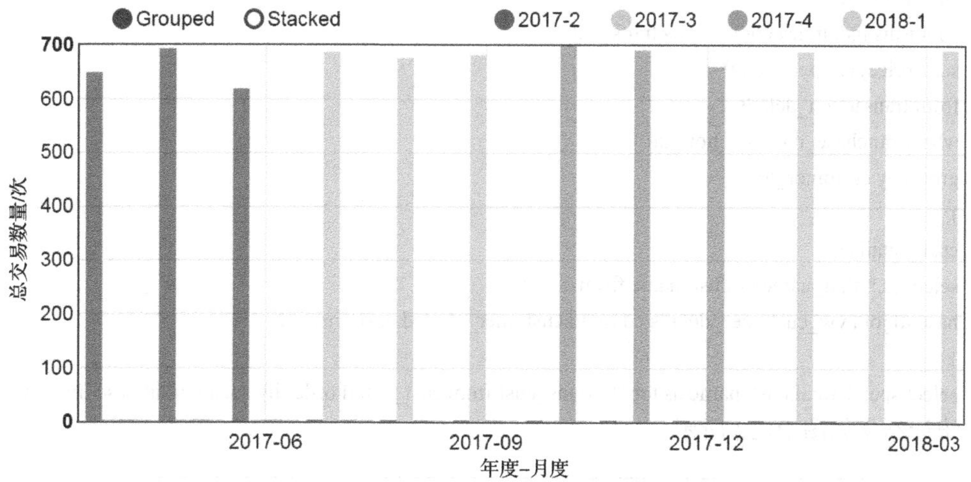

图10.20　年度-月度的总交易数量统计结果

⑧ 统计消费次数排行榜的前 10 位客户

关键代码：

```
%hive
with base as (
select customer_id,count(distinct transaction_id) as trans_cnt,sum(price) as spend_total
from transaction_details where purchase_month is not null group by customer_id),
cust_detail as (
select td.*,first_name as cust_name from
base td join vw_customer_details cd on td.customer_id = cd.customer_id)
select trans_cnt,cust_name as top10_trans_cust from cust_detail order by trans_cnt desc limit 10
```

统计结果如图 10.21 所示。

trans_cnt	top10_trans_cust
27	Cyrillus
27	Bondie
26	Ingeberg
26	Louis
26	Ashley
26	Tatum
26	Melania
26	Theodora

图10.21　消费次数排行榜的前10位客户统计结果

⑨ 统计消费额排名前 10 的客户

关键代码：

```
%hive
with base as (
```

```
select
customer_id,
count(distinct transaction_id) as trans_cnt,
sum(price) as spend_total
from transaction_details
where purchase_month is not null
group by customer_id
),
cust_detail as (
select td.*,first_name as cust_name from
base td join vw_customer_details cd on td.customer_id = cd.customer_id
)
select spend_total,cust_name as top10_trans_cust from cust_detail order by spend_total desc limit 10
```

统计结果如图 10.22 所示。

spend_total	top10_trans_cust
799.09	Tatum
787.14	Ashley
786.8	Louis
763.26	Alyson
755.39	Venus
743.19	Bondie
719.54	Luke
705.18	Theodora

图10.22　消费额排名前10的客户统计结果

⑩ 统计周期内消费次数最少的客户

关键代码：

```
with base as (select customer_id,count(distinct transaction_id) as trans_cnt
from transaction_details where purchase_month is not null group by customer_id)
select * from base order by trans_cnt limit 10
```

统计结果如图 10.23 所示。

base.customer_id	base.trans_cnt
133	5
240	6
263	8
329	8
194	8
279	8
110	8
370	8

图10.23　统计周期内消费次数最少的客户

⑪ 统计每年度-季度客户总数

关键代码：

```
%hive
with base as (select customer_id,
concat_ws('-', substr(purchase_date, 1, 4),
cast(ceil(month(purchase_date)/3.0) as string)) as year_quarter, substr(purchase_date, 1, 4) as year
from transaction_details where purchase_month is not null)
select   count(distinct customer_id) as total, year_quarter, year
from base group by year_quarter, year order by year_quarter
```

统计结果如图 10.24 所示。

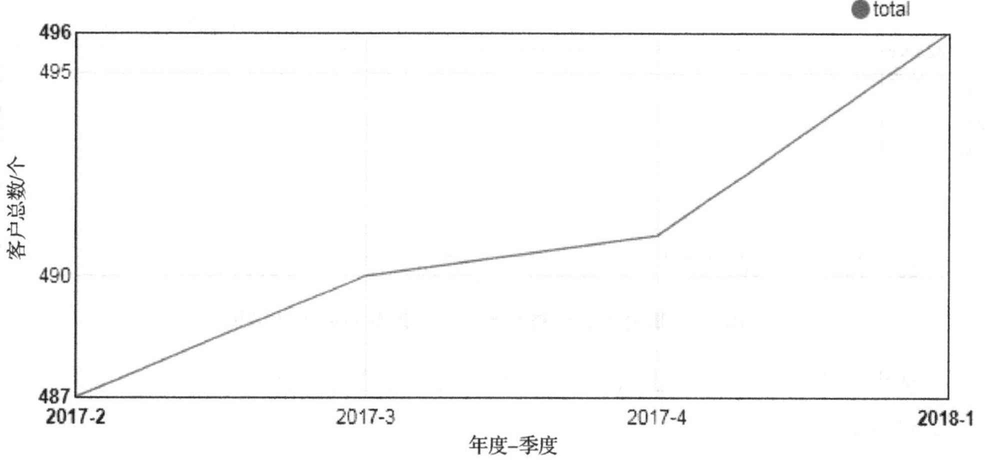

图10.24　每年度-季度客户总数统计结果

⑫ 统计最大的客户平均消费额

关键代码：

```
%hive
with base as (select customer_id,avg(price) as price_avg,max(price) as price_max
from transaction_details where purchase_month is not null group by customer_id )
select max(price_avg) from base
```

统计结果如图 10.25 所示。

_c0
36.671176

图10.25　最大的客户平均消费额统计结果

⑬ 统计每月最高消费额与最常来的客户

关键代码：

```
%hive
with base as (
select customer_id, purchase_month, sum(price) as price_sum, count(transaction_id) as trans_cnt
```

```
from transaction_details where purchase_month is not null group by purchase_month, customer_id),
rank_sum as (select
rank() over(partition by purchase_month order by price_sum desc) as rn_sum,
rank() over(partition by purchase_month order by trans_cnt desc) as rn_cnt,
purchase_month,price_sum,trans_cnt,customer_id from base)
select purchase_month,'spend' as measure_name,price_sum as measure_value,customer_id
from rank_sum where rn_sum = 1
union all
select purchase_month,'visit' as measure_name,trans_cnt as measure_value,customer_id
from rank_sum where rn_cnt = 1 order by measure_name, purchase_month
```

统计结果如图 10.26 所示。

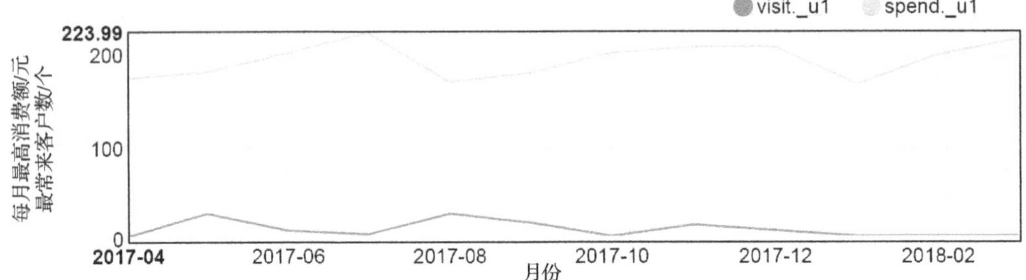

图10.26 每月最高消费额与最常来的客户数统计结果

⑭ 基于消费额统计受欢迎程度排名前 5 的商品并进行验证

关键代码：

```
%hive
select product,sum(price) as price_sum from transaction_details
where purchase_month is not null group by product order by price_sum desc limit 5
```

统计结果如图 10.27 所示。

product	price_sum
Goat - Whole Cut	311.55
Macaroons - Two Bite Choc	280
Dr. Pepper - 355ml	279.76
Nantucket - Pomegranate Pear	279.44
Wine - Merlot Vina Carmen	269.07

图10.27 基于消费额统计受欢迎程度排名前5的商品

⑮ 基于购买频次统计受欢迎程度排名前 5 的商品并进行验证

关键代码：

```
%hive
select product,count(transaction_id) as freq_buy from transaction_details
where purchase_month is not null group by product order by freq_buy desc limit 5
```

统计结果如图 10.28 所示。

⑯ 基于客户数量统计受欢迎程度排名前 5 的商品并进行验证

关键代码：

```
%hive
select product,count(customer_id) as freq_cust from transaction_details
where purchase_month is not null group by product order by freq_cust desc limit 5
```

product	freq_buy
Cheese - Camembert	11
7up Diet, 355 Ml	10
Goat - Whole Cut	10
Coriander - Seed	9
Dr. Pepper - 355ml	9

图10.28　基于购买频次统计受欢迎程度排名前5的商品

统计结果如图 10.29 所示。

product	freq_cust
Cheese - Camembert	11
7up Diet, 355 Ml	10
Goat - Whole Cut	10
Coriander - Seed	9
Dr. Pepper - 355ml	9

图10.29　基于客户数量统计受欢迎程度排名前5的商品

（3）门店分析

① 按客流量统计最受欢迎的门店

关键代码：

```
%hive
select sd.store_name,count(distinct customer_id) as unique_visit
from transaction_details td join ext_store_details sd on td.store_id = sd.store_id
group by store_name order by unique_visit desc limit 5
```

统计结果如图 10.30 所示。

sd.store_name	unique_visit
Walmart	486
Lablaws	478
FoodLovers	474
NoFrill	470
FoodMart	466

图10.30　按客流量统计最受欢迎的门店

② 按客户消费额统计最受欢迎的门店

关键代码：

```
%hive
select sd.store_name, sum(td.price) as total_revnue from
transaction_details td join ext_store_details sd on td.store_id = sd.store_id
group by store_name order by total_revnue desc limit 5
```

统计结果如图 10.31 所示。

sd.store_name	total_revnue
Walmart	44165.92
Lablaws	43279.59
FoodMart	41428.62
NoFrill	41291.93
FoodLovers	41092.24

图10.31 按客户消费额统计最受欢迎的门店

③ 按交易频次统计最受欢迎的门店

关键代码：

```
%hive
select sd.store_name,count(transaction_id) as unique_purchase
from transaction_details td join ext_store_details sd on td.store_id = sd.store_id
group by store_name order by unique_purchase desc limit 5
```

统计结果如图 10.32 所示。

sd.store_name	unique_purchase
Walmart	1680
Lablaws	1660
NoFrill	1595
FoodLovers	1584
FoodMart	1581

图10.32 按交易频次统计最受欢迎的门店

④ 按客流量统计每个门店最受欢迎的商品

关键代码：

```
%hive
with base as (select store_id,product,count(distinct customer_id) as freq_cust
from transaction_details where purchase_month is not null group by store_id, product),
prod_rank as (select store_id,product,freq_cust,
rank() over(partition by store_id order by freq_cust desc) as rn from base)
select store_name, product, freq_cust
from prod_rank td join ext_store_details sd on td.store_id = sd.store_id
```

```
where td.rn = 1
```

统计结果如图 10.33 所示。

store_name	product	freq_cust
Walmart	Aspic - Clear	4
Walmart	Whmis - Spray Bottle Trigger	4
Walmart	Pepper - Chilli Seeds Mild	4
Walmart	Onion - Dried	4
Walmart	Cake - Pancake	4
Walmart	Pork - Smoked Kassler	4
Walmart	Fruit Salad Deluxe	4
Walmart	External Supplier	4

图10.33　按客流量统计每个门店最受欢迎的商品

⑤ 统计每个门店客流量与雇员的比率

关键代码：

```
%hive
with base as (select store_id,count(distinct customer_id, purchase_date) as cust_visit
from transaction_details where purchase_month is not null group by store_id)
select store_name,cust_visit,employee_number,
round(cust_visit/employee_number,2) as cust_per_employee_within_period
from base td join ext_store_details sd on td.store_id = sd.store_id
```

统计结果如图 10.34 所示。

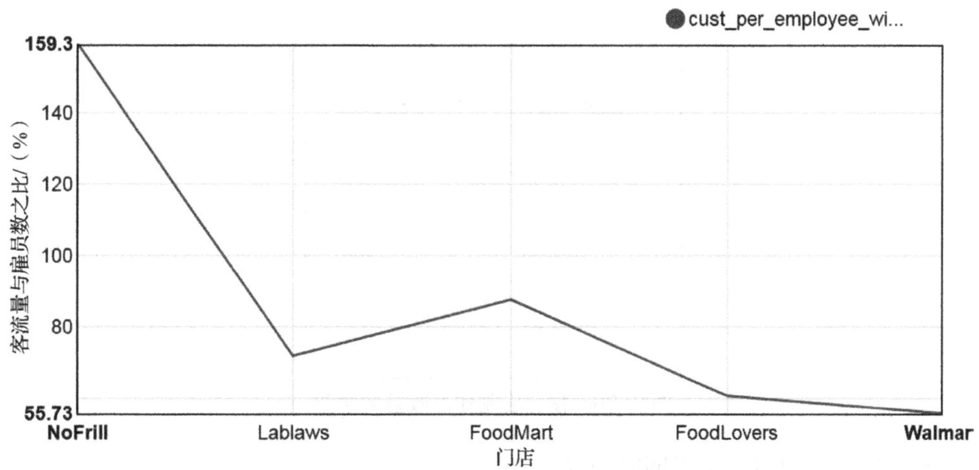

图10.34　统计每个门店客流量与雇员数的比率

⑥ 按年度-月份统计每个门店的收益

关键代码：

```
%hive
select store_name,purchase_month,sum(price) as revenue
from transaction_details td join ext_store_details sd on td.store_id = sd.store_id
```

```
where purchase_month is not null group by store_name, purchase_month
```

统计结果如图 10.35 所示。

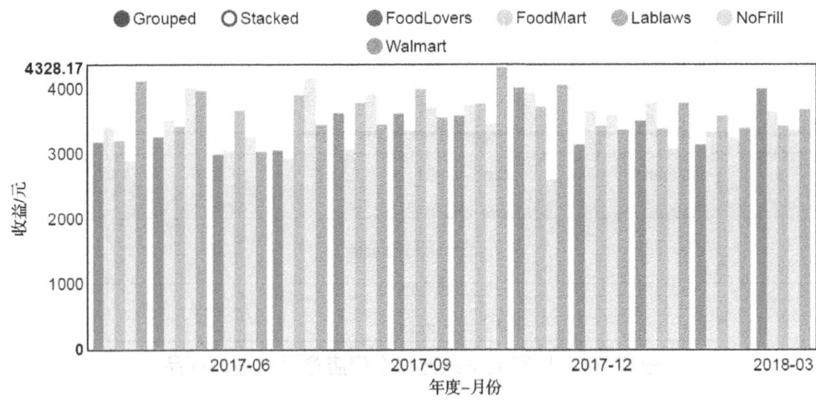

图10.35　按年度-月份统计每个门店的收益

⑦ 按门店制作总收益的饼图

关键代码：

```
%hive
select store_name,sum(price) as revenue
from transaction_details td join ext_store_details sd on td.store_id = sd.store_id
where purchase_month is not null group by store_name
```

代码运行结果如图 10.36 所示。

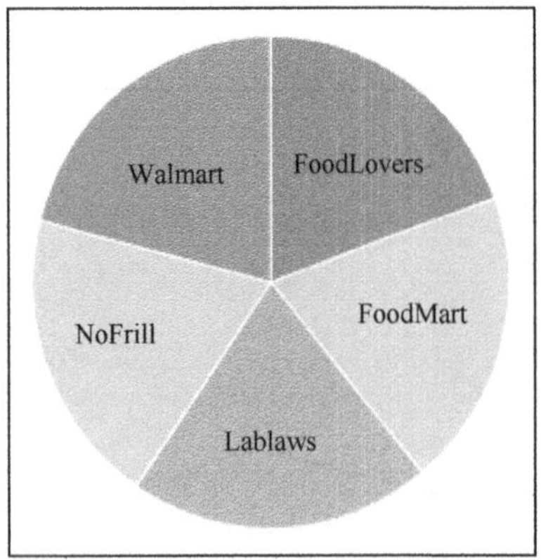

图10.36　按门店制作总收益的饼图

⑧ 统计每个门店最繁忙的时间段

关键代码：

```
%hive
with base as (
select transaction_id, purchase_time, if(purchase_time like '%PM',
concat_ws(':',string(hour(from_unixtime(unix_timestamp(purchase_time,'hh:mm')))+12),
string(minute(from_unixtime(unix_timestamp(purchase_time,'hh:mm'))))),
from_unixtime(unix_timestamp(purchase_time,'hh:mm'),'HH:mm')) as time_format,
```

```
store_id from transaction_details
where purchase_month is not null),
timeformat as (
select purchase_time, transaction_id,
(cast(split(time_format, ':')[0] as decimal(4,2)) +    cast(split(time_format, ':')[1] as decimal(4,2))/60)
as purchase_time_in_hrs, store_id from base),
timebucket as (
select transaction_id, purchase_time, purchase_time_in_hrs, store_id,
if(purchase_time_in_hrs > 5 and purchase_time_in_hrs <=8, 'early morning',
if(purchase_time_in_hrs > 8 and purchase_time_in_hrs <=11, 'morning',
if(purchase_time_in_hrs > 11 and purchase_time_in_hrs <=13, 'noon',
if(purchase_time_in_hrs > 13 and purchase_time_in_hrs <=18, 'afternoon',
if(purchase_time_in_hrs > 18 and purchase_time_in_hrs <=22, 'evening', 'night'))))) as time_bucket
from timeformat)
select sd.store_name, count(transaction_id) as tran_cnt, time_bucket
from timebucket td join ext_store_details sd on td.store_id = sd.store_id
group by sd.store_name, time_bucket order by sd.store_name, tran_cnt desc
```

统计结果如图 10.37 所示。

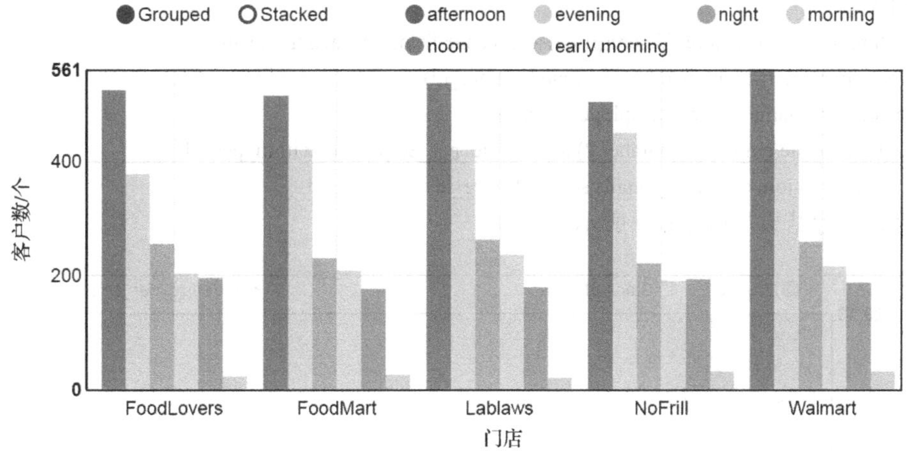

图10.37 统计每个门店最繁忙的时间段（通过客流量体现）

⑨ 统计每个门店的忠实顾客

忠实顾客指在统计周期内消费额排名前 5 的客户。

关键代码：

```
%hive
with base as (select store_name,customer_id,sum(td.price) as total_cust_purphase
from transaction_details td join ext_store_details sd on td.store_id = sd.store_id
where purchase_month is not null group by store_name, customer_id ),
rk_cust as (select store_name,customer_id,total_cust_purphase,
rank() over(partition by store_name order by total_cust_purphase desc) as rn
from base)
select * from rk_cust where rn <= 5
```

统计结果如图 10.38 所示。

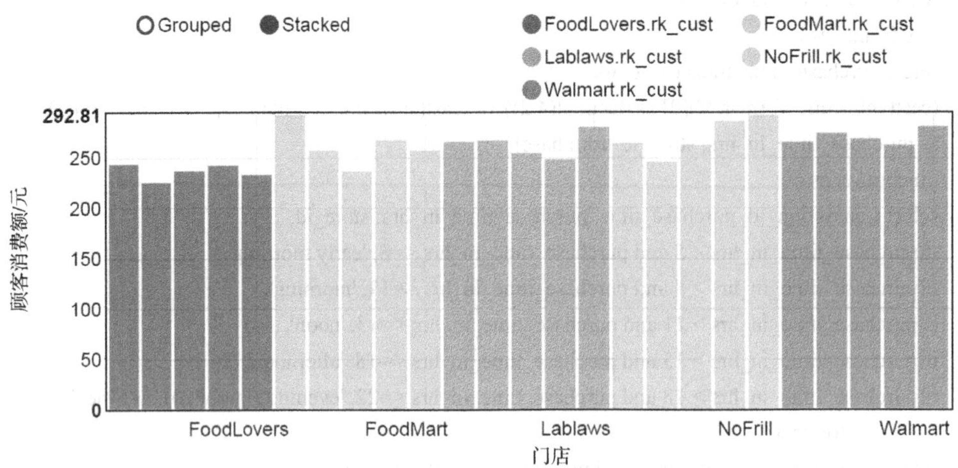

图10.38　统计每个门店的忠实顾客（通过顾客消费额确定）

⑩ 根据每位员工的最大收入确定明星店

关键代码：

```
%hive
with base as (select store_id,sum(price) as revenue from transaction_details
where purchase_month is not null group by store_id)
select store_name, revenue,employee_number,
round(revenue/employee_number,2) as revenue_per_employee_within_period
from base td join ext_store_details sd on td.store_id = sd.store_id
```

代码运行结果如图 10.39 所示。

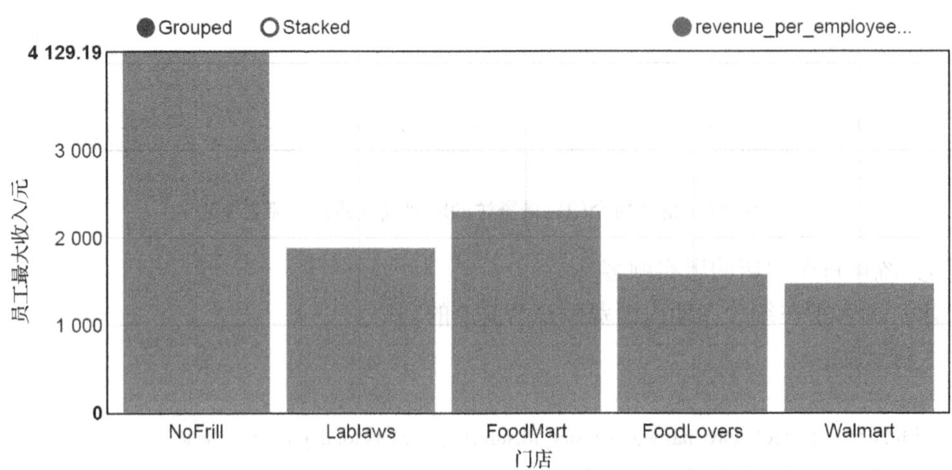

图10.39　根据每个员工的最大收入确定明星店

（4）评价分析

① 在原始数据表 ext_store_review 中找出存在冲突的交易映射关系数量

此处的“冲突”是指“transaction_id”与“store_id”组合键在交易表和评价表中不同。

关键代码：

```
%hive
select count(*) from
transaction_details td join ext_store_review sd on td.transaction_id = sd.transaction_id
where purchase_month is not null and td.store_id != sd.store_id
```

代码运行结果如图 10.40 所示。

_c0
725

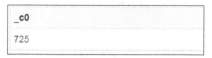

图10.40　评价表中存在冲突的交易映射关系数量

② 找出客户评价覆盖范围

关键代码：

```
%hive
select count(td.transaction_id) as total_trans,
sum(if(sd.transaction_id is null, 1, 0)) as total_review_missed,
sum(if(sd.transaction_id is not null, 1, 0)) as total_review_exist
from transaction_details td left join ext_store_review sd on td.transaction_id = sd.transaction_id
where purchase_month is not null
```

代码运行结果如图 10.41 所示。

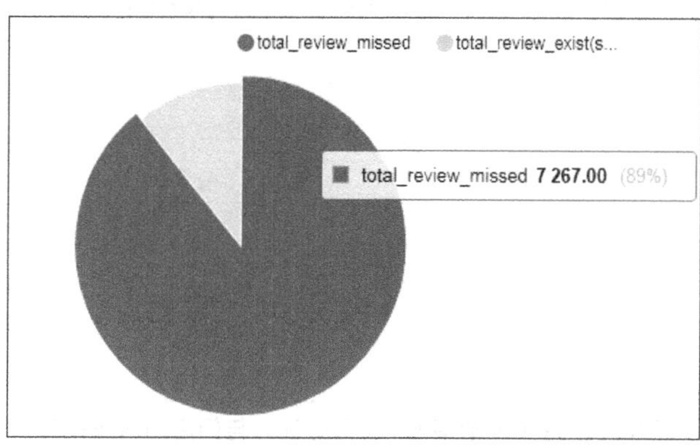

图10.41　客户评价覆盖范围

③ 根据评分了解客户与交易的分布情况

关键代码：

```
%hive
select review_score,count(distinct customer_id) as num_customer,count(*) as num_reviews
from transaction_details td join ext_store_review sd on td.transaction_id = sd.transaction_id
where purchase_month is not null and review_score <> '' group by review_score
```

代码运行结果如图 10.42 所示。

④ 客户是否总是给同一家门店最佳评价（5 分好评）

关键代码：

```
%hive
select count(*) as visit_cnt,customer_id,td.store_id
from transaction_details td join ext_store_review sd on td.transaction_id = sd.transaction_id
where purchase_month is not null and review_score = '5'
group by customer_id, td.store_id order by visit_cnt desc
```

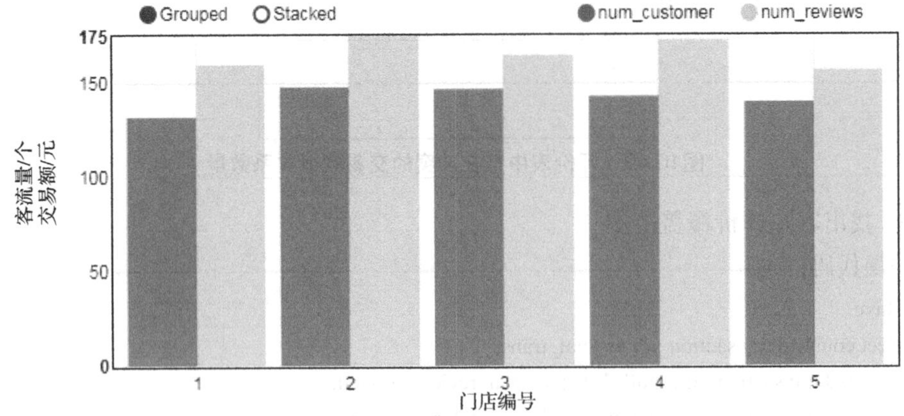

图10.42　根据评分了解客户与交易的分布情况

代码运行结果如图 10.43 所示。

visit_cnt	customer_id	td.store_id
2	239	3
2	28	3
2	491	1
1	99	3
1	93	5
1	93	2
1	92	5
1	9	4

图10.43　客户最佳评价（5分好评）与门店的对应关系

本章小结

通过完成“电子商务消费行为分析”项目实训，读者不仅能够对使用 Hive 进行数据分析和处理有更深刻的理解，还能够熟练使用 Hive 进行数据变换与查询，进一步加强对可视化工具的使用。

本章作业

独立完成“电子商务消费行为分析”项目。